Introduction to Bioethics

D1449506

Introduction to Bioethics

John Bryant
Professor, School of Biological Sciences University of Exeter

Linda Baggott la Velle
Senior Lecturer, Graduate School of Education University of Bristol

John Searle
Exeter and District Hospice; Anglican Priest

John Wiley & Sons, Ltd

Other Wiley Editorial Offices

John Wiley & Sons Inc., 111 River Street, Hoboken, NJ 07030, USA

Jossey-Bass, 989 Market Street, San Francisco, CA 94103-1741, USA

Wiley-VCH Verlag GmbH, Boschstrasse 12, D-69469 Weinheim, Germany

John Wiley & Sons Australia Ltd, 33 Park Road, Milton, Queensland 4064, Australia

John Wiley & Sons (Asia) Pte Ltd, 2 Clementi Loop #02-01, Jin Xing Distripark, Singapore
129809

John Wiley & Sons Canada Ltd, 22 Worcester Road, Etobicoke, Ontario, Canada M9W 1L1

Wiley also publishes its books in a variety of electronic formats. Some content that appears in
print may not be available in electronic books.

Library of Congress Cataloging-in-Publication Data

British Library Cataloguing in Publication Data

A catalogue record for this book is available from the British Library

ISBN 0 470021977 Hardback
     0 470021985 Paperback

Typeset in SNP Best-set Typesetter Ltd., Hong Kong
Printed and bound in Great Britain by Antony Rowe Ltd., Chippenham, Wilts
This book is printed on acid-free paper responsibly manufactured from sustainable forestry in
which at least two trees are planted for each one used for paper production.

Contents

Preface

Each new power won by man is a power over man as well. Each advance leaves him weaker as well as stronger. In every victory, besides being the general who triumphs, he is also the prisoner who follows the triumphal car.

A recent comment, we may think, that captures the concerns that many have about how our recently acquired knowledge in biomedical science will be used. But actually these words were written by the British scholar of mediaeval English, C.S. Lewis, in his 1947 book *The Abolition of Man*. Jumping forward 55 years to 2002, we find very similar views expressed, albeit in more 21st century language, by Francis Fukuyama in his book *Our Post-Human Future*. Of course, there has always been concern about the ways in which humans use the knowledge they have gained. However, in the areas of biology and medicine, the possibilities for misuse seem to some to be more sinister or more personally threatening than in the past. This is coupled, in a post-modern social climate, with at times a growing mistrust of science and a greater willingness to question the direction, application and even sometimes the findings of science. At the same time, progress in biological science, biomedical science and medicine is happening very fast. Indeed, it is so rapid that many are caught off-guard when it comes to debating the social and ethical issues that are raised by this progress.

It was against this background that in 1995/96 we introduced a bioethics course for undergraduate students in Biological Sciences and in Science Education at the University of Exeter. Our reasoning was that, given the pace and nature of scientific advance, the ethical issues it raises and the prevailing social climate in which it takes place, scientists need the intellectual tools to be able to engage in the ethical debate. As we stated in the preface to a previous book,[1] *too much of the debate, especially in the media, is conducted with little scientific understanding. Scientists who recognise and understand the ethical dimension can make a major contribution.* The Exeter bioethics course has been very successful and has been adopted as a model in universities elsewhere. Indeed, there is a growing national (within the UK) and international impetus for teaching ethics to bioscience students. It is the existence of such courses that provided some of the reason for writing this book. In

[1] *Bioethics for Scientists* (2002) eds Bryant, J., Baggott la Velle, L. and Searle, J. Wiley, Chichester.

producing the previous book, *Bioethics for Scientists*, we were aiming mainly at those who would lecture on such courses, giving a strong scientific as well as ethical and philosophical grounding. In writing this current volume our major objective has been to provide a text mainly for the students, especially students of biology, medicine and other life sciences who have a good background in the science but who are less knowledgeable in ethical theory and moral philosophy. We have been very aware of student needs as we have written this book; we very much hope that it meets those needs and indeed we thank those students who have told us what they would like to see in a readily available text book.

We are grateful to many friends and colleagues for good discussion of these issues and for the way that they have helped (and sometimes challenged!) our thinking, especially Alex Aylward, Celia Deane-Drummond, Steve Hughes, Nancy Jones, Suzi Leather, Stephen Minger, Joe Perry, Christopher Southgate, Philippa Taylor, Peter Turnpenny and also Chris Willmott, Andy Bond and other members of the Higher Education Academy's Special Interest Group on Teaching Ethics to Bioscience students. We acknowledge that we have been particularly helped by the writings of Christopher Southgate on environmental ethics and of Christopher Hooks and Philippa Taylor on nanotechnology and cybernetics. We thank Claire Alexander and Sue Baxendale for advice on interpretation of the Human Tissue Act and Susan Pickering for allowing us to use her photographs of early embryos, embryo biopsy and human stem cells. Last but not least, we also want to thank our colleagues at John Wiley and Sons, especially Andy Slade, who has supported and encouraged us throughout the project, and Robert Hambrook who took the book through all its production stages.

John Bryant
Linda Baggott la Velle
John Searle

Exeter and Bristol, June 2005

1 Science and society

I feel the story should be told, partly because many of my scientific friends have expressed curiosity about how the double helix was found, and for them an incomplete version is better than none. But even more important, I believe, there remains a general ignorance about how science is 'done'. That is not to say that all science is done in the manner described here. This is far from the case, for styles of scientific research vary almost as much as human personalities. On the other hand, I do not believe that the way DNA came out constitutes an odd exception to a scientific world complicated by the contradictory pulls of ambition and a sense of fair play.

From *The Double Helix*, James D Watson (1968)

1.1 What's it all about?

This is a book about bioethics but we are starting with a consideration of the practice of science and its relationship with wider society. Why? Consider the four following case studies.

Case study 1

- *Donated gametes – sperm and ova – are used in fertility treatments for patients who are unable to produce their own.*

- *It is much easier, for obvious reasons, to donate sperm than ova.*

- *Donated ova are very scarce.*

Introduction to Bioethics, by John Bryant, Linda Baggott la Velle and John Searle
Copyright © 2005 by John Wiley & Sons, Ltd.

- *During foetal development, females lay down a lifetime's supply of oocytes (egg cells).*

- *It has therefore been suggested that aborted female foetuses may be used to supply oocytes/ova for fertility treatments.*

Do you approve or disapprove of this idea? What are your reasons?

Case study 2*

- *A small less-developed country in South America is deep in debt.*

- *Its main resource is its rain forest.*

- *In order to raise revenue, the government has granted a licence to a Japanese logging company to clear 25% of the forest.*

- *The land that has so far been cleared is used for cattle ranching, mainly to raise beef for the US market.*

- *The government has also granted a licence to a trans-national biotechnology company to exploit the forest's gene pool.*

- *In addition to the income from the licence, the company has agreed to pay royalties on income generated from discoveries based on the rain forest gene pool.*

What are the issues involved in dealing with this situation?

* Based on a study presented by Southgate, C.C.B. (2002) in *Bioethics for Scientists*, eds Bryant, J., Baggott la Velle, L. and Searle, J. Wiley, Chichester, pp 57–72.

Case study 3

- On several occasions recently, normally fertile couples have sought permission to undergo in vitro fertilization in order to produce a baby that can be a stem cell donor for an older sibling.

- In several of the cases, the older sibling suffers from a genetic disorder, and the embryos created in vitro would be tested for the absence of the mutation and for a positive tissue match to the older sibling.

- In another case, the condition suffered by the older sibling is not 'genetic' but the child still needs donated stem cells. In this case the in vitro embryo would be selected solely as a tissue match.

In which of these cases, if any, would you grant permission? Give your reasons.

Case study 4

- A small biotechnology company in Mexico has discovered a gene that encodes a protein in the network of resistance to oxidative stress in plants.

- Laboratory experiments have shown that when the gene is transferred by genetic modification techniques to crop species, the crop plants show an enhanced capacity to grow and produce yield under conditions where water supply is limiting.

- The company has not published its data because it is filing a patent on the gene.

- If the patent is granted, the company plans to license it out to a major trans-national agri-chemical company.

Should the patent be granted? Give your reasons.

These case studies are on the surface very different from each other. However, they all describe situations in which ethical dilemmas have been raised by advances in science and by the way that the science, through its application, may have impacts on the lives of individuals and/or on wider society. The issues presented in these case studies are discussed in detail in later chapters. In the meantime it is important to consider briefly the factors that influence our decision-making in these and similar situations.

• First, there may be an immediate personal reaction – a 'gut response' – along the lines of 'Yuk, that's awful' or 'Wow, that's brilliant' or along more socio-political lines – 'That's just not fair/not right'.

• Second, there will (it is hoped) be a more thought-out ethical analysis which may complement the gut reaction but which may also cause the gut reaction to be questioned.

• Third, it is important to realize that both gut response and the more thought-out ethical analysis are very likely to be affected by one's world-view and personal philosophy (which for some will include religious commitment).

• Fourth, because advances in science are embedded in all these studies, one's view of the science itself is important. Do we know all that we need to know in order to go ahead or is more work needed? Are the conclusions presented in support of a particular proposal soundly drawn? Do these scientists know what they are doing? Should the basic research have been done in the first place?

Thus, science is one of the factors that informs bioethical decision-making; we cannot avoid thinking about science, why and how it is done and how it relates to wider society. And that is what we explore in the rest of this chapter.

1.2 What is science?

Introduction – some history (but not very much)

We get the word *science* from a Latin word, *scio*, which means 'I know', and in the original usage science simply meant knowledge. The application specifically to knowledge about the material nature of the universe, gained by a particular set of methods, dates back less than 200 years. Some of those whom we regard as the great scientists of the past, such as Isaac Newton, would not have called themselves scientists. Indeed, Newton's position at Cambridge was Lucasian Professor of Mathematics and his major work was called (translating from the Latin original) *The Mathematical Principles of Natural Philosophy*. The latter term effectively means physics and was used in this way in the older Scottish universities well into the second half of the 20th century. However, we now have a very clear idea of what we mean by the more general term science: the word implies a whole approach to the material world, to methods of acquiring knowledge about that world and to the body of knowledge thus acquired. So, how did we arrive at this situation?

To an early human being, the world around must have seemed a strange and often hostile place. It was certainly a place of contrasts, embodying both provision and threat. So while plants could be harvested, some were poisonous; while animals could be hunted, some animals, including some quarry animals, were very dangerous. Further, there were (and indeed still are) unpredictable and often devastating events such as storms, earthquakes and volcanic eruptions. Nature was not to be taken lightly and it was important that knowledge of the positive and negative aspects of the natural world was passed on verbally from generation to generation. Doubtless humankind's investigation and knowledge of nature remained at this level for tens of thousands of years. However, dating from about 75 000 years ago there is evidence of art, and as that art over successive millennia became more sophisticated it relied on quite detailed observations of nature. One just has to look at rock art and cave paintings in places as diverse as Australia, France, Siberia, South Africa[1] and Spain dating from between 25 000 and 10 000 years ago to become aware of this. Furthermore, as cultures evolved, so did descriptive knowledge of the times and seasons, so that there was confidence that the sun would rise daily and that the seasonal rains would fall, that certain animals migrated and that plants grew at particular times. Some of this knowledge may have been very sophisticated; in Britain, for example, the alignment with the sunrise on the summer solstice and the sunset on the winter solstice of particular stones in the stone circle at Stonehenge indicates quite a detailed knowledge of astronomical events through the year. Stonehenge dates at about 2800 BC, around the same time as the period of building pyramids at Giza in Egypt was under way. The alignment of the pyramids shows that the Egyptians could ascertain the direction of true north, another indication of growing knowledge of the natural world.

The Egyptians also put knowledge about the natural world into good use in their daily lives. The river Nile provides water in a land which would otherwise be very arid. The ancient Egyptians observed that the river flooded every spring and that the silt spread by the floods provided a fertile substrate for growth of crops. Indeed, by measuring the volume of the flood water at different places, estimates were made of the likely crop yield that year (and therefore what the tax 'take' was likely to be!). However, despite this sophis-

[1] The oldest known art, dated at about 77 000 years before the present, is a carving on a haematite nodule, found in South Africa. Beads made from ostrich egg shells dating back about 38 000 years have been found in the same region.

tication, there was apparently no knowledge of the spring run-off from the mountains of the upper Nile basin, which causes the annual flooding. The Egyptians were thus observers of nature as it affected their lives, but was this science? They also applied their observations and had significant engineering expertise, expertise good enough for the building of the pyramids, but again we may ask whether it was science.

To the extent that science simply equals knowledge, the ancient Egyptians (and the ancient Britons who built Stonehenge) were scientists, but as far as we can tell there was no theorizing about the reasons for the phenomena they observed, beyond ascribing them to the work of myriad gods. It was in the Greek culture, with its emphasis on mind, that theorizing about the reason for and the nature of the universe began to flourish, and this theorizing was tied in with other areas of thought, including especially mathematics, philosophy and ethics (see Chapter 2). The Greeks, like the Egyptians, were accomplished builders and technicians, putting their knowledge to practical use, but they were certainly not great experimenters, despite Archimedes' fortuitous bath-time discovery about volume and water displacement from which he cleverly deduced information on the density of metals. So, although the flowering of Greek culture saw the development of theories about many natural phenomena, even a great physician such as Hippocrates carried out very little actual experimental testing of the theories. So again, to the extent that the Greeks added to our knowledge of the universe they practised science. However, there was no universal view that theories should actually be tested by experimentation.

This latter point is emphasized because we now regard the testing of hypotheses as one of the major ways in which science advances. The knowledge is obtained not just by observation and theorizing but by making further observations and by carrying out experiments. In western culture this understanding of science dates back to Francis Bacon (1561–1626), who, impressed by the discoveries made by Copernicus, insisted that understanding nature required evidence that could only be gathered by experiment, by careful measurement and by rigorous observation. This change of emphasis is known as the Baconian revolution, and Bacon is often referred to as the *father of science* and the *secretary of nature*. He and his contemporary Galileo are credited with abolishing for ever the Aristotelian view of nature. The adoption of Bacon's approach led to a rapid expansion of scientific knowledge in the 17th and 18th centuries, typified by, for example, the work of Newton, and leading thence to modern science.

1.3 Modern science

Science as practised in the 21st century continues to embody the principles set out by Bacon and thus we can say that science is an investigation of the material nature of the universe by a set of methods that include observation, experiment, the erection of hypotheses and the testing of these hypotheses, but within this overall definition there is room for much variety. Different sciences place different emphases on observation and experiment. Hypotheses come in different forms as do methods of testing them. Science as practised is not a single type of activity, although it all takes place within a single over-arching framework. This was clearly understood by Nobel laureate James Watson, whose words head this chapter.

Let us then open this up a little more and explore briefly some ideas in the philosophy of science and the nature of scientific knowledge. This is important because misunderstandings of what science is and how it works can lead to negative attitudes to science, to scientists and to the applications of science. A most important basic principle is that, at any one moment, scientific knowledge is incomplete (we do not and cannot know everything) and provisional (it is possible that our current understanding may be modified by subsequent findings). For this reason many aspects of scientific 'knowledge' are actually the hypotheses that are open to further testing. Nevertheless, scientists assume that there is an objective reality to which this partial and provisional knowledge relates. This is what the science philosopher Polyani calls *verisimilitude* – approach to the truth. Progress in scientific knowledge and understanding is generally said to made by the 'scientific method' that was outlined above and in particular in the testing of hypotheses. Further, the science philosopher Karl Popper maintained that 'real' hypotheses are those for which there is the possibility of being proved wrong (i.e. are falsifiable). Therefore, according to this view, science can only progress by the formation of falsifiable hypotheses, which are then tested by further work. This seems a very sterile description of an activity that many find very exciting.

Indeed, amongst many practising scientists and growing numbers of science philosophers, there is a view that the 'Popperian' approach to science is too sterile and stereotyped. Science is actually more flexible. It embodies serendipity (making significant discoveries by accident), intuition (in which an interpretative leap is made that goes beyond the strict limits of what the data tell us) and even guesswork. When Watson and Crick turned one strand of the double helix upside down (and in doing so achieved a workable and essentially correct model for the structure of DNA) they were acting on either a 'lucky' guess or a piece of brilliant intuition, depending on who one reads. So science can make progress by methods other than the direct testing of specific hypotheses, although of course these 'non-conventional' findings can themselves be verified or falsified by subsequent work, as in the double helix, where the opposite orientation of the two strands was confirmed by experiment.

The strictly conventional view of science also fails on two other grounds. First, it is clear that scientific hypotheses come in a variety of forms; some are very well established and are so widely and generally applicable that they should be regarded as paradigms. Evolution comes into this category. Indeed, the use of the term 'theory' or 'hypothesis of evolution' has led to a good deal of misunderstanding amongst those who seek to promote other views. Other hypotheses are very local in application and may also be very tentative because of the scarcity of relevant information such as when we have data based on observations of just a tiny number of patients, or from one small experiment. Secondly, Popper's description of real hypotheses as those being those capable of being proved wrong cannot be universally applied. Experiments are often carried out in order to ascertain whether there is evidence to *support* rather than refute a hypothesis. Further, there are some facets of scientific knowledge which, as pointed out by John Polkinghorne,[2] are here to stay; these include atoms and the helical structure of DNA. Popper's view of science does not accommodate gains in knowledge.

1.4 Science, ethics and values

Science progresses in a step-wise manner; some of the steps are large (and then the public media often talk of a breakthrough) but mostly they are small. But whether the steps are large or small, whether the new data support or refute an earlier hypothesis, one thing is clear: science progress depends on what has gone before. If one of us sets up an experiment that is based on published data it is expected that those data were not falsified or fudged and that the author in whose paper the data appear has given a correct version of what he or she has done. We can only see further than previous scientists because we are, metaphorically, standing on their shoulders (whether or not they are giants). The reader will be quick to appreciate that this implies a trust in those who have gone before, a trust that they did not make up their data. Without this ability to trust what other scientists publish, the whole edifice of science would tumble. A parallel situation occurs in competitive sport where throwing a game for the sake of financial reward is seen as going against the whole ethos of sporting competition. Thus, amongst other responsibilities (as discussed in Chapter 13), a scientist has ethical responsibilities to the whole science community, indeed to science itself. To suggest that a scientist has lied about his or her results (as has happened in some of the debates about particular applications of science) is a very serious accusation. Indeed, scientists who are discovered to have in any way presented false data are generally dismissed from their posts, as has been evident from a handful of high-profile cases in the past ten years.

In addition to the ethics specifically associated with the practice of science, we must also emphasize that the science is not value free. The impression of the scientist working in a social vacuum, driven just by curiosity, is no longer

---

[2] Quoted by Bryant and Searle in *Life in Our Hands*, (2004) IVP, Leicester.

valid and perhaps never was. At the personal level, scientists may speak of competition, of the race to reach a particular research goal and of the desire for having one's name associated with a major discovery. James Watson suggests that he and Francis Crick selected the structure of DNA because it was then the biggest prize in biological science. Personal ambition is often a major driver of the scientific enterprise but more altruistic motives may also lead to research on particular topics; for example, some are drawn to work on vaccines for malaria or on drought-tolerant crops because they hope for applications that will aid less-developed countries. The scientist does not leave behind his or her aspirations, world-view or personality when entering the laboratory. Indeed, the latter may affect the choice of research area and the context in which the research is performed.

For scientific discovery there is an important parallel here with learning theory in general. The influential Russian psychologist Vygotsky wrote of the importance of the 'zone of proximal development', meaning that the social and physical environment is vital for learning to take place. He believed that a successful learner is in some manner 'scaffolded' – supported – by 'able others'. Whilst scientists pushing forward the frontiers of our understanding can often make intuitive deductive leaps based on the interpretation of their observations, this individual effort is often, in terms of the overall investigation, a small part of the whole, and others will have significantly contributed to that breakthrough moment.

Therefore, the context in which science is done is socially constructed. The gentleman- or lady-scientist, doing original research paid for from their own financial means, is a person of the past. Science has grown into a major world activity, embedded into national economies and employing across the world many tens of thousands of people. In the developed world, the applications of science are woven into our daily lives, and are very much taken for granted. Science publishing is now a major business, with thousands of journals competing with each other to attract the best research papers in their particular subject area. Modern science needs extensive funds and the allocation of funds for particular types of research is a societal decision, whether made as a result of government policy or of industrial priorities. Even in so-called blue skies research it is easier to obtain funds for some research topics than for others. Resource allocation reflects what society at the time deems to be valuable.

Case study

You are the head of a university biology department. The university promotions committee has asked you to nominate one and only one of your academic staff (faculty in US ter-

minology) for promotion. There are two obvious possible candidates.

Candidate A is 37 and is very highly respected internationally for his work on the ecology of plant–insect relations. His research on the evolution of pollination mechanisms is widely respected as is his knowledge of plant and insect communities in the Peruvian Andes. The research has received a steady flow of grant funding from government funding agencies.

Candidate B is 34 and is building up a strong reputation for her research on the regulation of gene expression in programmed cell death, especially in relation to cancer. Her recent papers on the switch between cell 'immortality' genes and cell death genes in mice have caused great interest in the biomedical community and have been widely quoted. The work is supported by extensive grant funding from government agencies and from medical charities and this high level of funding has led to her having one of the larger research groups in the department.

Which candidate do you select and why?

Science and values – some more examples to ponder

- *Some forms of human cancer may be studied by inducing their formation in genetically modified mice.*

- *Francis Crick claims that he switched from physics to biology with the intention to abolish the last vestiges of vitalism from the latter science.*

- *Radioactive isotopes are used in research, in diagnosis and in some medical treatments. A knock-on effect of these activities is the discharge into the environment, under strictly regulated conditions, of radioactive material.*

- *Francis Collins, US director of the Human Genome Project, agrees with Copernicus that investigating and understanding nature is one of the highest forms of worship of God.*

- *Richard Dawkins, Professor of Public Understanding of Science at Oxford University, believes that science will eradicate what he calls the superstition and fantasy of religion.*

- *Genetic testing of an individual may reveal information that could, if divulged to an employer, be disadvantageous to that individual.*

So then it is clear that there are ethical issues arising from some types of scientific research. These include the use of animals, possible environmental damage, the participation of human subjects, concerns about possible applications of results, and allocation of scarce (financial) resources, to mention a few. There are also issues relating to individual and to societal values. We cannot say that science is value free, although some scientists still try to do so. All these have a bearing on the way that science is regarded and in the way that its findings are applied. We therefore continue by examining the changing attitudes to science.

1.5 Attitudes to science

Science and the Enlightenment

Societal attitudes to science in the early years of the 21st century are somewhat different from those of 40 years or so ago. A closer look at changes in prevailing world-views shows why this may have occurred, especially in northern Europe. The Baconian revolution in science occurred very early in a period characterized by an intellectual movement known as the Enlightenment which, from roots in the 16th and 17th centuries, flourished especially in the 18th century on both sides of the Atlantic. The Enlightenment placed great value on the abilities of humankind; the Church was no longer seen as the source of all knowledge. The use of human reason was regarded as the major way to combat ignorance and superstition and to build a better world. Many of the adherents of the Enlightenment movement rejected religion and were humanists. On the other hand, there were also Enlightenment thinkers who did not reject religion and regarded the human mind as the pinnacle of God's creation. Thus, whether religious or not, members of the Enlightenment movement placed great stress on the human intellect. Combining this with the Baconian approach to investigating nature thus placed science in very high esteem.

Science, modernism and post-modernism

Although the Enlightenment as a movement died out towards the end of the 18th century, many of its attitudes continued into the 19th century,[3] including for the most part a positive attitude to science and its applications. There were however, some voices of dissent, early signs of an arts–science divide. Goethe suggested that the view of the world espoused by Newton and his successors was cold, hard and materialistic, turning nature into a machine. The romantic poet Keats, referring to Newton's work, wrote

> *Philosophy will clip an Angel's wings,*
> *Conquer all mysteries by rule and line,*
> *Empty the haunted air, and gnomed mine –*
> *Unweave a rainbow . . .*

However, in general, the 19th century witnessed widespread applications, especially of the physical sciences, in technology and engineering. There was continued confidence that science could reveal objective truth about the world and that human ingenuity could put that knowledge to good use. Thus emerged a philosophy known as modernism, which, although we can trace its beginnings back through the enlightenment to the Baconian revolution, flourished in the later years of the 19th century right through into the middle years of the 20th century. Therefore, although for many the occurrence of two world wars dented idealistic views of humans as moral agents, there remained an immense confidence in humankind's creative and technological abilities. In 1964, Harold Wilson, then the prime minister of the UK, spoke of the country benefiting from the 'white heat of technology'. Science and technology shaped many aspects of culture in the 1960s on both sides of the Atlantic. The contraceptive pill opened the way for a widespread change in sexual behaviour; there was great public interest in the conquest of space; telecommunications and information technology were on the verge of huge expansion. The press (but not the science community) spoke of nuclear energy as likely to provide 'electricity too cheap to meter' thus providing an 'atoms

---

[3] Indeed, in ethics, 20th century initiatives to define a basic set of human rights (as discussed in the next chapter) can be traced back to Enlightenment thinking.

for peace' counter to background angst about nuclear warfare. Such confidence in science in all its aspects continued in general right through the 1970s.

However, the arts–science divide that surfaced early in the 19th century was becoming more marked. The scientist, public administrator and novelist C. P. Snow wrote extensively in his novels about the work of scientists in public life and about the relationship between science and other aspects of society and culture. In 1959, he coined the term 'the two cultures' to describe, in the educated classes in the UK, a great divide between science and the arts. His claim was that, despite the central position of science and technology in modern life, a high proportion of well educated people understood very little about science, a cultural divide that continues today. Further, we also need to note that by the 1970s a philosophical shift had already started, a shift towards post-modernism.

In order to understand this philosophical shift, we need to look back to the 19th century philosopher Nietzsche. Based on his view that 'God is dead', he suggested that there are no external reference points; individuals define for themselves their own moral and cultural values and indeed are free to 're-invent' themselves. This leads to a fragmentation in ideas about truth and culture. If individuals can define their own moral values, then there is nothing to stop a person deciding on courses of action that work out best for themselves, rather then having wider terms of reference. This approach to moral decision-making is known as rational egoism and is the most extreme of the consequentialist ethical systems (as discussed in the next chapter), in that it considers only the consequences for the individual making the decision. It is thus in the philosophy of Nietzsche that we see the origins of post-modernism, a belief that anyone's world-view, concept or version of the truth or ethical value system is as valid as anyone else's. If this leads an individual to adopt rational egoism as an ethical system, so be it.

Although the roots of post-modernism were planted in the 19th century, its growth and flowering have been very much a feature of the 20th and early 21st centuries. It is not our intention here to discuss this philosophy in detail, but we do need to mention some of the main strands within it. In the UK, the Cambridge philosopher Wittgenstein insisted that words, including scientific terms, must be interpreted in their social context. This, taken to its ultimate conclusion, leads to the view that no word can have a universally accepted meaning,[4] that there can be no underlying universal truth, a conclusion that is certainly reached by writers such as Derrida and Foucault and the 'deconstructionist' school of literary criticism, all of whom emphasized the absence of universal truths, of over-riding themes or 'meta-narratives'. In some academic circles it is now acceptable to state that 'all things are rela-

---

[4] We are reminded of a conversation between Alice and Humpty Dumpty in Lewis Caroll's *Alice Through the Looking Glass*: 'When I use a word,' Humpty Dumpty said, in a rather scornful tone, 'it means just what I choose it to mean, neither more nor less'.

tive', despite the inherently self-defeating nature of this statement; relativism has thus become a distinct philosophy under the post-modern umbrella.

Although the average person in the street probably has not heard of post-modernism, this mode of thinking has certainly seeped into popular culture, especially in northern Europe (rather less so in the USA). Indeed, as we have suggested elsewhere,[5] it is probable that most people in the UK, especially in the under-55 age-group, think in a post-modernist way, very much influenced by the media, which have been pervaded by post-modernism. An over-arching post-modernism will clearly affect general ethical thinking, as mentioned above and as discussed in the next chapter, but what about science?

If all 'truth' is culturally constructed, then this will include scientific truth, so post-modernism will say that published scientific data have little or no relation to objective reality, even if it is accepted that the scientists themselves have published those data in good faith. In the most extreme versions of this view, it is suggested that the actual results obtained by the scientists are socially constructed. Obviously, if this were so, the whole edifice of science would collapse. Experiments done in one continent within one culture would yield different results from the same experiments done in another continent within another culture. That is not the experience of the scientific community and scientists in general have not espoused post-modernism, at least in respect of science. Post-modernism is thus seen as a threat to science. However, scientists do acknowledge that because science is an activity of people, its practice is not free from personal values, including reasons for choosing particular lines of research, personal ambition, altruism, desire for recognition and so on. Science is not done by robots. Further, in the practice of modern science, some types of research are regarded as more deserving (or demanding) of financial support than others; there is thus, as we noted earlier, a strong societal element in the support of science. However, scientists argue strongly that the actual results obtained in scientific experiments are not socially constructed. The source of the money does not determine the outcome of the research. Nevertheless, it is acknowledged that there are cases in which results may have been suppressed because of commercial interests, as happened, for example, in the tobacco industry with data indicating the adverse effects on health of smoking.

Finally, we return to public attitudes to science. There is certainly now a greater ambivalence towards science than in the middle years of the 20th century, at least in northern Europe. So, although the influence of science and technology is as central as ever, it is not uncommon to hear anti-science views expressed, some of which echo the words of Goethe and Keats. The seepage of post-modernism into general modes of thinking has affected societal attitudes to science, although rather less so in the USA than in Europe. Thus, an individual may accept or reject a particular scientific finding according to

[5] Bryant, J. and Searle, J. (2004) *Life in Our Hands*, IVP, Leicester.

whether it coincides with that individual's pre-existing ideas or, indeed, whether it useful to do so. The discovery of a gene involved in a particular disease may be hailed as breakthrough: the application of science in medicine generates a good deal of respect for the authority of science. On the other hand, the findings of some scientific enquiries are rejected because they do not coincide with the views of particular groups. In post-modernism, my view is as valid as yours, even if you are the so-called expert. Thus, acceptance of scientific authority and of the validity of scientific findings is patchy; this in turn has an effect on reactions to the ethical issues that arise from science, the issues that form the subject matter of the rest of this book.

Case study: employing science to sell a product, the modernist and post-modernist versions

In the 1960s and 1970s it was not uncommon for TV advertisements to involve a white-coated, usually male, scientist figure to validate particular products such as pain-killers or detergents (as the Rolling Stones sang in Satisfaction: *'I am watching my TV and a man comes on and tells me how white my shirts can be').*

In marked contrast, consider the following, typifying advertisements for shampoo on UK TV in 2004. A female actor or voice-over tells viewers that hair is 96% amino acids. It is then stated that the shampoo in question is 'rich in aminos' and thus using it can nourish the hair, replace lost 'aminos' and generally improve hair health.

Analysis:

1   Yes, hair is 96% amino acids but these amino acids are chemically joined together in a long protein chain called keratin. Therefore, a part-truth has been used to set the scene.

2   Amino acids cannot be joined into protein molecules except by the process of protein synthesis. A protein cannot be repaired or replenished by direct uptake of amino acids.

3   In any case, the hair takes up only a very small pro-
      portion of the amino acids from the shampoo.

4   The process of protein synthesis that makes the keratin
      takes place in the hair cell at the base of the hair and
      not in the hair itself.

5   If shampoo entered the hair cell in anything but the
      most minute quantities, the detergent in the shampoo
      would disrupt many cellular processes including protein
      synthesis. So the shampoo cannot deliver amino acids
      to the site of protein synthesis (and the hair cell has its
      own supply anyway).

6   Note the use of the term 'amino'* rather than amino
      acid – the term acid is dropped as soon as possible
      because of the negative connotations it carries.

This is not a comment about the efficacy of any particular
shampoo; we are sure that modern shampoos clean the hair
and the scalp and leave the hair shiny and manageable.
Rather, it is a comment on the dishonest use of scientific terminology to imply things that cannot happen. Essentially, the
advertisers are saying that if they want to use the jargon of
science to sell their product, they will do so on their terms.
It is classic post-modern triumph of style over substance.

* In fact, 'aminos' are currently one of the major themes in shampoo advertis-
ing. In the USA, for example, it is claimed for one shampoo that it 'adds aminos
that build protein that your hair needs'.

2 Ethics and bioethics

Nothing is wrong, nothing is right. Energy is pure delight ... everything that lives is holy

Final message from *Jerry Springer – The Opera*, Stewart Lee and Richard Thomas (2003)

What rubbish this is

Comment on this final message by William Rees-Mogg in *The Times* (10 January 2005) following a screening of *Jerry Springer – The Opera* on BBC TV.

2.1 Introduction

This chapter explains what ethics and bioethics are, using the following headings:

- What is ethics?
- The development of ethics
- The growth of bioethics
- Ethics in the 21st century
- Making decisions.

2.2 What is ethics?

Telling lies is wrong. Our relationships with each other only function well if there is a presumption that what we say to each other is true. Trust is essential in human relationships and in public life. One of the most painful

Introduction to Bioethics, by John Bryant, Linda Baggott la Velle and John Searle
Copyright © 2005 by John Wiley & Sons, Ltd.

experiences for parents is if they discover that one of their children has lied to them. Similarly, children's confidence in adults and indeed themselves can be seriously harmed if their parents lie to them. Politicians tend to lose elections when they lose the trust of the electorate. But is telling lies always wrong? Suppose you are a German living in Berlin in the Second World War and that you are hiding a Jewish family in your cellar. One day, the Gestapo arrive at your house and ask 'Are there any Jews around here?'. Almost certainly your answer will be 'No'. You lie in order to protect those you are hiding. You recognize that telling lies is usually wrong, but under these circumstances surrendering Jewish fugitives to the gas chambers is worse.

While ethics is about what we ought and ought not to do, it is also about setting priorities in human behaviour. Ethics is not always about what is absolutely right or wrong, acceptable or unacceptable, ideal or less than ideal. It is also about what is the best decision in particular circumstance, what is the lesser of two evils, what is the balance between doing good and causing harm. Ethics is therefore about working out the principles on which we make these sorts of decisions. That is what this chapter is about. In the rest of the book we apply these principles to the complex and exciting developments which are taking place so rapidly in biological and biomedical science.

First, however, it is important to understand the relationship between law and ethics. Ethics forms the foundation on which law is built but not all ethics is enshrined in law. Murder is wrong. It is unlawful to do it and if I do and am found guilty I will be punished. Here the ethical principle of the uniqueness and preciousness of human life is enshrined in law. However, the law recognizes that I may kill another person under differing circumstances and adjusts the penalty for doing so accordingly. The side of the road on which we drive is not itself a question of right or wrong. But as there would be a serious risk of injury or death to ourselves and others if we chose individually on which side to drive, the law decides whether we drive on the right or on the left. The law also intervenes where there is a conflict between individuals about the best interests of other people, so the courts often have to decide which parent should have the custody of children when a couple divorce. Parents and doctors may disagree about whether or not a child should have medical treatment. Quite properly, the courts are asked to analyse the ethical principles in each case, set out what the law says and decide what is in the best interests of the child.

However, there are innumerable matters where we make our own judgements. The England rugby player, Jonny Wilkinson, refuses to be photographed without his shirt on. He is not only anxious to protect his privacy but also believes that it is more appealing if a guy keeps his shirt on. The England football[1] captain, David Beckham, takes a different view and is more

---

[1] Soccer to our American readers.

than happy to be photographed without his shirt. They have different views about privacy and sex appeal.

2.3 The development of ethics

Introduction

The development of ethical thinking or moral reasoning is a long and complex story in which many different strands intertwine, fall apart and are reconfigured. Religious and non-religious thinkers have been engaged in this process for at least 3000 years. Debates have often been fierce. At times, decisions and practices have been driven less by objective reasoning and more by events. Some of the views put forward by the great thinkers of the past are difficult for us to grasp and seem very odd to us in the 21st century. In our own society there is a great divide between religious and secular views. However, the long story of the development of moral reasoning continues to influence the decisions we make now. The contributors to that story are very many indeed. Here we shall only mention some of its most influential.

The ancient Greeks

Part of the story begins in ancient Greece where the authoritative source of moral reasoning was the epic poems of Homer, put together in the 8th century BCE. Epics such as the Trojan War were about courage, justice, heroism, piety, lust, love and the relationships between humans and the gods. These were what guided people. However, Socrates (470–399 BC/BCE) questioned the usefulness of these stories, asking what the good life really was. As a result he was accused of corrupting the youth of Athens, condemned and executed by poison. In accepting his sentence he enunciated a important principle, 'It is better to suffer wrong than to do wrong'.

Socrates' student, Plato (427–347 BCE), wrote up much of what Socrates had said and developed his own complex theories of ethics. It is from Plato that much later the Church began to develop its ideas of dualism, separating out the body from the spirit, the material world from the spiritual world, the former of these often being considered flawed and less important than the latter. Plato's theory of 'forms' is difficult for those of us brought up in a scientific world to understand. One example may help. He thought that things such as beauty or goodness had their own forms which existed in their own world. What we see and experience are, as it were, reflections of these forms. We can only see and experience them fully when the soul separates from the body. The human task is to reflect on them and so participate in them in some way. Plato also extended these ideas to objects, so that a chair was but an expression of some form of a chair existing in its own world. When this sort of thinking was taken up by the Church it had far reaching effects on its ideas about the human body and the world in which we live. Coupled with its view that the body or the 'flesh' was something bad, it is easy to see how, for example, sex came to be seen as something rather smutty.

In the next generation of Greek thinkers was Aristotle (384–322 BCE), himself a student of Plato. He moved away from Plato's ideas of 'forms'. His main concern was to find out what was the chief human good. He understood that such good was achievable but also realized that people disagreed about what constituted good and happiness, so he linked happiness to function. Function is about expressing virtue or excellence. We determine what excellence is as the soul listens to reason. Reason shows us a middle way between human excess and deficiency. (So, for example, greed is an excess but starvation is a deficiency. No doubt Aristotle would have considered that neither the grossly obese nor those with an eating disorder had listened to reason. Excellence or virtue in these areas would be about sensible eating habits.) Good is about building excellence into our lives. Excellence is then dependent on reason. Indeed Aristotle believed that the highest good is about the exercise of reason in study and contemplation.

Judaeo-Christian ethics

While the Greeks were grappling with these issues the Hebrew Bible (what we know as the Old Testament) was being put together. Central to Jewish ethical thinking was the belief that God had spoken through the ancient patriarchs (Abraham, Isaac and Jacob), the lawgiver Moses and the prophets who were active over several centuries. At the heart of what God had said were the Ten Commandments and the various codes of conduct about religious practice, hygiene, personal morality and the activities of the state. When these were observed there was individual and national prosperity. It was the responsibility of the prophets to tell people when they and the community were disobeying them and warn them of the dire consequences that would follow unless they mended their ways. Indeed, the Old Testament history of the Jewish people is written in this context.

Following the life of Jesus Christ, the Christian Church developed its own literature of the four gospel accounts of Jesus, letters from church leaders to churches throughout the Roman Empire, an account of how some these churches were planted and predictions about the end of time. This is the New Testament, which was put together with the Old Testament to form the Bible, to become not only a text of religious doctrine and practice but also of ethical teaching. One of the main tasks facing the early Christians was how to work out their ethics in the world of the Roman Empire, of Greek thought, in which there were many gods and in which Christians were persecuted.

Everything changed during the reign of the Emperor Constantine (306–337 AD/CE). Christianity became the formal religion of the Empire. From then on, as Europe was christianized, the ethics of western civilization was moulded by the Church, and its leaders became increasingly powerful not only in the Church but also in the State. There were many downsides to this, not least when the Church became oppressive and abused its power. Nonetheless, much morality and law-making was based on the view that there were God-given absolutes about how individuals and communities should conduct themselves. Inevitably though, philosophical thought, experience and later science impacted on these, so that within the Church debates developed about what was absolute and could not be changed and what should change in the light of human progress and development. That debate continues today for example over human sexuality. Is it moral for two people of the same gender to enter into an intended life-long relationship, not just of friendship but also of sexual intimacy? (Of course, on this particular issue, western society has largely moved on and for many such debates are irrelevant.) It is to these developments in philosophy and science that we must now turn briefly.

Natural law

The theory of natural law goes back to the 5th century BCE. The Stoics believed that a good life was a life lived in accordance with nature, not just on a human level but also in accordance with the natural order of the world of which human beings are part. In developing this, Aristotle believed that nothing in nature was produced without having a purpose, that purpose being for everything to be itself as fully as possible. Trying to work out what this fullness is was the study of 'forms'. For human beings, reason and contemplation were the highest 'form'. As this theory of living in accordance with the natural order at the fullest level developed, it was inevitable that it should be increasingly rooted in scientific and philosophical reflection, so from what science and philosophy tell us we can work out moral values and obligations. However, where scientists, philosophers and indeed anyone differ in their views about human nature there will be different views about moral values and obligations. For example, are human beings primarily motivated by desires to experience pleasure and avoid pain, or are we rational beings who make decisions on the basis of reasoned consideration rather than our desires?

Thomas Aquinas (1225–1274) developed natural law theory. He concluded that as we reflect on our human nature we come to see to what ends or purpose we are naturally inclined. He started with Aristotle's idea that human beings have a function. He broke this down into the functions of the different parts of the body, moving from general function to specific function. We function best if we have good health, good education and the freedom to choose. To deny these things to people is therefore immoral. This became the basis of much Roman Catholic ethical teaching and indeed remains so. However, when it is applied, for example, to the area of sex, it has caused much dissension. In Thomastic thinking (i.e., thinking based on the work of Thomas Aquinas) the function of sex is to produce children. Therefore, any sexual activity where the intention is not to have children or where the possibility of conception is prevented by using birth control is immoral. Any idea that sex may be for pleasure is simply not allowed. A further weakness in natural law is that it assumes we will always come to the most moral answer as to what we should do. Human experience shows that this is often not the case. Self-interest easily intrudes into our decision-making.

Ethics after the 16th century

The birth of modern science through the work of Copernicus (1473–1543), Galileo (1564–1642) and Sir Isaac Newton (1642–1727) seriously undermined religious concepts of the universe (even though all three of these were Christian believers). Formerly, the universe was conceived of in biblical terms – the heavens above, the Earth beneath and the waters under the Earth. It was a three-decker arrangement, the centre of which was the Earth and around which the sun moved. When such ideas were shown not to be true, it was inevitable that the religious basis of ethics would be questioned also. Was morality something determined by God or was it something we had to work out for ourselves? The 18th century, particularly, saw the development of rationalism, the concept that knowledge and eventually ethics must be based on rational principles.

A key figure in 18th century ethical thinking was Immanual Kant (1724–1804). He was professor of logic and metaphysics at the University of Konigsberg in East Prussia. He was not satisfied with a system of ethics based on God's revelation. He believed that it was only reason that could legislate in a dispassionate and universal manner. From this he developed his 'imperatives', the basis on which human beings ought to act. The most influential was that we should 'act so as to treat humanity never only as a means but also as an end'. In other words, the end does not justify the means (a crucial area of debate in modern biomedical ethics). More difficult to grasp is Kant's 'categorical imperative'. This was his supreme principle of duty. It is not about the ends achieved by an action. Rather it is about principles that a rational person would act on. These principles are universal. If a principle cannot be universally applied we ought not to act on it. For example, we help people in need because if we think that there are circumstances in which we may legitimately fail to provide such help, we might one day find ourselves in need, but other people, adopting our principle, would fail to help us. There is, therefore, a categorical imperative always to help people whom we know to be in need. What perhaps Kant failed to address is what we do when our imperatives clash with each other – as we have seen in the case of telling lies. Nevertheless, Kant has been hugely influential in forming concepts of duties, autonomy and universal ethics.

Consequentialism

Another ethical theory that developed in the 18th and 19th centuries was consequentialism. This says that the rightness or wrongness of an action ultimately depends on its consequences. The dominant version of this is utilitarianism, particularly associated with Jeremy Bentham (1748–1832) and John Stuart Mill (1806–1873). Right actions are those that produce the greatest amount of pleasure for those affected by its consequences. Conversely, something is wrong if fails to generate pleasure but rather produces pain or harm. What is right is that which maximizes good outcomes, the most good for the most number of people. Bentham developed a system of calculating the pleasure/pain balance. While such an attempt seems distinctly odd to us in the 21st century, the principle continues to the major determinant in framing public policy and making political decisions. For example, the British Prime Minister, Tony Blair took the view that it was better both for the people of Iraq and the security of the West for Saddam Hussein to be removed. This 'good' outweighed the violence and chaos that Saddam's removal unleashed. Others, of course, took precisely the opposite view. As we shall see, utilitarianism has become an important principle in deciding what should and should not be allowed in modern biomedical research and practice.

In summary, by the end of the 18th century there were two main ways of approaching ethics. First there was *deontological* ethics, based on absolute values predominantly derived from religion (at least in Judaeo-Christian thinking). Something was either intrinsically right or intrinsically wrong and people had an overwhelming obligation (duty, as in the Greek word *deon* meaning duty) to pursue the former. Attempts had also been made to try and work out such absolutes from reason without using religion as the primary source for coming to a particular view, for example in the work of Kant. Second there was *consequentialism*, particularly as expressed in utilitarianism. What mattered was the outcome of a decision. There was also a third approach, known as *virtue ethics*, which we discuss later on in this chapter.

2.4 The growth of bioethics

During the 25 years after the Second World War (1939–1945), several factors came together to give rise to the birth of the discipline of bioethics. These were

- the rapid advances in biomedical science

- the perceived inadequacy of traditional ethics

- the Nuremberg war crime trials

- decreasing paternalism and deference

- concern for the environment.

A frog had been cloned in Cambridge in the early 1950s (see Chapter 9). In 1953, Watson and Crick made their ground-breaking discovery of the structure of DNA. Nevertheless, when the authors of this book were studying biology at A-level,[2] our courses in genetics were mainly concerned with the process of cell division and patterns of inheritance. The structure of DNA was being taught in the senior years of secondary school but much was still based on classical genetics. It was not until the early 1970s that modern molecular biology really took off, opening up our later ability to manipulate genes, to study them in great detail, and to diagnose and select against genetic disease in the early human embryo. Massive advances were also taking place in many other branches of biomedicine: human organs were being transplanted; life could be prolonged by drugs and surgery; the function of organs such as the kidneys and the lungs could be taken over by machines. Studies of mammalian fertilization led to the creation of 'test-tube' babies. Genetic tests could be applied to embryos created in vitro. Mammals were eventually cloned.

So enormous were the possibilities raised by these advances that there was increasing concern in society that the traditional framework of ethical thinking could not bear their weight. How could religious absolutes be applied to issues about which such texts as the Bible and the Koran knew nothing? It was much more difficult to weigh the balance between benefit and harm (in Bentham's language, pleasure and pain). What for example was the benefit of keeping somebody alive for many weeks by machines only to find that their subsequent quality of life was severely reduced? Furthermore, traditional ethical principles conflicted at times. Should the early human embryo be given all the legal protection owed to a human being, thereby preventing research being carried out in certain areas of disease when such diseases could be eliminated by such research?

A further concern was the degree of scrutiny of scientists and doctors by society. This had its origins in the Nuremberg trials after the second World War, where the actions of scientists, doctors, lawyers, military personnel and politicians were subject to forensic investigation and criminal proceedings by

---

[2] A word of explanation for our readers outside the UK: A-levels (i.e. Advanced level school certificate examinations) are public examinations taken in the last year of secondary school in England and Wales. Acceptance into university depends on the grades obtained at A-level.

the Allies. In many cases individuals were judged to have acted outrageously, violating the most basic respect for their fellow humans. As the 20th century moved on, the public demanded to have more say in how biomedical discoveries were used. Sometimes this was part of the decline in paternalism and deference that was taking place in Western society as it became more egalitarian. At other times there was frank mistrust of scientists. So bioethics developed as a discipline, notably in the United States of America. In bioethics, philosophers, lawyers, theologians, sociologists and lay people join with biomedical scientists in assessing what is the appropriate use of new developments and technologies. In many areas there are now formally constituted groups, which have a major input into public policy and the regulation of science.

However, many of these issues were about individual choices: my desire to have a child; my wish not to have a handicapped child; my anxiety about being kept alive only to face a poor quality of life. Much wider concerns were developing in the early 1960s and 1970s, particularly about the environment. Hitherto, the assumption had been that the Earth and its resources were primarily for the benefit of human beings, with little thought being given to the effects such use would have on the environment. However, as early as the 1940s concerns were being expressed about human encroachment on the wild places of the world and the damage this was causing, and then, in 1962, Rachel Carson's seminal book, *Silent Spring*, drew attention to widespread and often deeply damaging chemical pollution. It was a cell biologist, V. R. Potter, who brought these concerns together in his book *Bioethics – Bridge to the Future*, published in 1970, and he is credited with coining the word *bioethics*.

Thus bioethics embraces not just the effects of scientific advances on individuals but also on communities, the environment and non-human species. Environmental ethics, genetically modified crops and animal welfare are all part of bioethical discussions. However, during this period of rapid change in the second half of the 20th century huge advances were taking place not only in biomedical science but also in philosophy and ethics. The principles of both were being radically reviewed so that the framework for ethical making in Western society at the beginning of the 21st century is by and large very different from that at the beginning of the 20th century. It is to this that we must now turn.

2.5 Ethics in the 21st century

Introduction

From the 17th century onwards, the main planks of ethical decision-making in the Western world were the Judaeo-Christian religion, philosophy and knowledge. Different individuals and groups gave more importance to one or other of these but even the non-religious recognized that in matters of law and public policy the Ten Commandments, the teaching of Jesus Christ and the moral codes of the Bible were hugely influential. During the 20th century, many other factors became important in making ethical decisions. These factors fall under two broad headings:

- the nature of the issues

- the nature of society.

The issues

Many of the issues on which decisions have to be made in the 21st century were unknown even 30 years ago. Since that time very difficult questions have been raised because so much has become possible. Here are three examples.

Questions

- *Is a person on a life support machine who has been deeply unconscious for days, with no evidence of recovery, alive, or is dying being prolonged artificially? Do you keep the life support machine going or turn it off?*

- *If an early human embryo has a genetic defect that will cause cancer in adult life, should it be destroyed or be allowed to develop into a baby?*

- *Because bombs and missiles can be targeted precisely should governments more readily embark on war to remove a cruel regime, as civilian casualties can be minimized?*

In each of these examples there is the potential to confer benefit and to cause harm. The answer therefore has to balance these. The weight given to the various factors involved in arriving at that balance will vary depending on the viewpoint of those making the decisions.

Society

There are four important features of Western society that have emerged over the last 50 years, which have been very influential in ethical thinking. First, Western society is pluralist. There are many different religions and philosophical frameworks to which people look for guidance. Often people look to several of them, taking a viewpoint from one and adopting a practice from another. This is brought sharply into focus by the increasing emphasis on individualism. Personal decisions are less about what effect the decision might have on other people and more about what an individual wants. A good example of this is the practice of voluntary euthanasia in Holland (see Chapter 12). The view there is that if I find my illness intolerable and there is little or no prospect of recovery, I have an individual right to have my life ended. This is now lawful practice in Holland. However, it has also resulted in involuntary euthanasia where people's lives may be ended not because they think their suffering is intolerable but because somebody else thinks so.

Second, much of Western society is post-Christian. Although in England Anglican bishops continue to sit in the House of Lords (at the time of writing), adherence to the Christian faith and regular church-going have declined massively in Western Europe. While many sociological studies show that people continue to regard 'spirituality' as important, they have almost no knowledge or understanding of the Christian faith. Many people see this as a good thing as they perceive Christian morality as being restrictive, shackling human freedom rather than enhancing it (I want to sleep with anyone I fancy – it's fun, but the Church says that sex must be confined to marriage – that is just old fashioned). Indeed the Church is often regarded as irrelevant to modern society.

Third, Western society is post-modern. Post-modernism is a patchwork of philosophies, which began to take recognizable shape in the 1970s. Essentially, it denies that there are universal values. Human beings do not have any external point of reference. Everything, including science must be interpreted in its social context. There is no such thing as objective truth be it philosophical, religious or scientific. Everything is relative. What matters is what works for me. It is therefore unacceptable for one person to insist that I adopt their beliefs or patterns of behaviour. Post-modernism is evident in many parts of society, particularly in the media and in some aspects of education. Scientific education and research, however, remain firmly based on hypothesis, experimentation and rational conclusions. Indeed one of the difficulties in the 21st century is that many people, because their thinking has been influenced strongly by post-modernism, simply do not accept scientific conclusions or recognize the validity of the process by which they are reached.

Fourth, the concept of *rights* has become very influential. This is not a new idea. It is enshrined in the American Constitution, where citizens are declared to have the 'right to life, liberty and the pursuit of happiness' and in the cries of the French Revolution of liberty, equality and fraternity. These right and fundamental concepts have evolved over the succeeding two centuries into a mass of rights and demands. For example: a woman has a right to an abortion; I have a right to a child; when a mistake is made or an accident occurs, I have a right to compensation. Historically, the enshrining of rights in constitutions and conventions was to prevent the abuse of people by those with power over them. This is exemplified by the concept of human rights enshrined in the United Nations *Charter of Human Rights*, which we may regard as a worthy effort to ensure universal standards for the treatment of other humans in a pluralistic world. However, it is doubtful whether rights alone are an adequate basis for ethics without corresponding responsibilities and duties. Thus I may have a right to a child, but I also have a responsibility to care for it and bring it up. I may have a right to recreational sex, but I also have a responsibility to ensure that an unwanted pregnancy does not occur or that a sexually transmitted disease is not spread by my activities. The pursuit of rights alone often infringes the rights of others.

2.6 Making ethical decisions

So far we have seen that ethical decision making has a long history, arises from different religious and philosophical positions, has many strands and has become increasing complex. However, the simple fact is that ethics is about making decisions, about making choices; do I do this or that, and these choices have consequences. Actually, we engage in ethical decision making every day of our lives.

- I pass a man in the street selling *The Big Issue*.[3] Do I buy one or not?

- I am running behind schedule and may be late for an important meeting. Do I break the speed limit and arrive on time or keep to the speed limit and arrive late for the meeting?

- I have been given a piece of English prose by my teacher to translate into French at home. Do I do it unaided or do I type the English into my computer and find an Internet programme which will translate into French for me?

- I am allowed to claim up to a given sum of money for expenses. Do I adjust my expenses upwards or claim for what I have actually spent?

---

[3] A magazine sold in the UK by homeless people as part of a programme to help them get their lives together.

Even our everyday conversation has an ethical component to it. When I am recounting something that happened to me do I tell it accurately or embellish it just a little to impress people?

So how do we make these day to day decisions as well as deciding on the big issues currently in the public domain? An approach that has become popular in recent years is *virtue ethics*, to which we referred earlier in this chapter. This asks 'what is the most virtuous, best or right decision?'. It is actually not a modern concept at all, going back to Aristotle and Plato. It is a continuous thread through Judaeo-Christian ethics and in the thinking of 18th and 19th century philosophers. It has several components. An important one is that we become virtuous by practising virtue. Thus I can truly only be an honest person if I practice honesty and so I become trustworthy. Virtue is more about the expression of character than about keeping rules. Others see virtue ethics as the framework within which moral judgements are made. Nevertheless, virtue is itself a relative concept. What I consider to be virtuous depends on my belief system. A very good example is the position of the Christian moral right in the United States. It sees abortion and same-sex civil partnerships as un-virtuous or immoral, but it is apparently held as virtuous or moral to embark on a war against Iraq, hold suspected terrorists outside the due processes of law and abandon international agreements on the environment.

So we are still faced with the question how do we make ethical decisions? There are several systems to help us to this. They all require us to think clearly about the options available to us; what are the consequences of this decision or that? What are the benefits or harm which might follow? Who might be harmed or helped? What are the premises from which I am coming? A very helpful approach has been provided by two American ethicists, Tom Beauchamp and James Childress, in their book *Principles of Biomedical Ethics*, first published in 1979. As the title suggests, they were particularly concerned with the rapid advances in biomedical science, but their framework is useful in many other areas too. It recognizes that there are several ethical principles that have to be taken into account, that they have to be prioritized, that they do not always have the same weight and that at times they conflict.

The first principle is autonomy, that is to say we may not do something against a person's wishes. Thus a doctor may not embark on medical treatment against a patient's wishes, however necessary the doctor may consider such treatment to be. However virtuous it may be to give money to charity, my employer cannot take a percentage of my salary and donate it to charity for me. Autonomy of course has its limitations. I may wish to smoke in a public place but is my autonomy in this matter outweighed by the harm that passive smoking may do to other people? This leads to the second and third principles.

Our responsibility to other people is to benefit them. We should not harm others. Therefore, a doctor's duty is to provide beneficial treatment not harmful treatment. In practice, of course, this is often balancing act. A drug, for example, may be beneficial but it inevitably has side-effects. Therefore, the decision to prescribe it is about weighing these two things up. Does the benefit outweigh the harm? Can the harm be limited? If we liberalize gambling laws, people who wish to gamble may more easily do so and government revenue is increased, but there is a risk that more people will become addicted to gambling, with all the consequences this has for them and their families.

The fourth principle is that of justice. That is to say, we have a responsibility to look to the wider consequences of our decisions. Again, this is easily understood in medicine. A new drug may be beneficial to a small number of patients but it is very expensive. In England, funds for healthcare are limited and therefore to spend money on the drug may mean that money is less available for things which would benefit many more people. So what does a hospital do – spend money on the drug or replace the chairs in the old people's unit?

These principles have been have been criticized as being but a pallid version of traditional ethics. Certainly they are based on key principles, not least from the Judaeo-Christian tradition and Greek and Enlightenment/humanist principles of seeking what is the most good and of not using people as means to an end. Their value is that they provide a framework for thinking, of exploring all the options and consequences of a proposed action. Particularly in the area of modern biomedical science, it is all too easy to respond to a development with either 'Yuk' or 'Wow'. Our gut reaction may be that this particular development is terrible; it must never be allowed to happen. Or it may be this is wonderful; it opens up so many possibilities. These were precisely the responses to the birth in 1978 of Louise Brown, the world's first test tube baby. Human sperm was used to fertilize a human egg in the laboratory where the embryo was allowed to develop for a few days. It was then implanted into the mother's uterus. The baby was carried by her for a normal pregnancy and Louise was born. Inevitably this involved several embryos being produced in the laboratory but only one becoming a baby. Some people said 'This is terrible. We are playing God. Embryos, which are human beings, are being destroyed in the process. Furthermore, you should not separate reproduction from sexual love. Yuk!'. Others thought it was wonderful. 'Now couples who cannot have a child by natural means will be able to have a family. Wow!' While both responses are understandable, neither thinks through all the issues involved. Ethical decision-making demands careful, thoughtful reflection if we are to make the best virtuous decision we can.

Elsewhere we have set out a stepwise process for doing this.[4] A similar approach has been proposed recently by the American bioethicists Adil Shamoo and David Resnik;[5] it is worth quoting albeit in a slightly modified form:

[4] Bryant, J. and Searle, J. (2004) *Life in Our Hands*, IVP, Leicester.
[5] Shamoo, A.E. and Resnik, D.B. (2003) *Responsible Conduct of Research*, Oxford University Press, New York.

The problem: Alice is faced with giving blood or honouring an obligation to attend a committee meeting. If she gives blood she will be benefiting others. Failure to do so would deprive somebody of that benefit. If she goes and gives blood she will be breaking a promise to attend the meeting. What should she do?

Here are the steps for Alice to take.

1 State or define the problem. Alice's problem is 'should I give blood or go to the meeting?'.

2 Gather relevant information. What is the need for blood? Is there a need for donors of her particular blood group? What is the nature of the meeting? How important is it that she be there?

3 Work out the options. Are these mutually exclusive? Could she go to the meeting and give blood on another occasion? Could the meeting time be changed?

4 Relate the options to the principles at stake. Giving blood supports the principle of helping others. Going to the meeting supports the importance of keeping your word. It may be possible to meet both principles by changing the times of one or the other. There may be other principles that also have to be considered – legal, economic, professional or personal.

5 Take everything into account. Here facts are important as well as principles. Alice may have a rare blood group and blood of that group is urgently needed; these facts should carry considerable weight in making the decision. She may have a common blood group and the meeting is the only one at which she can make a presentation about a new development in the industry that could be very worthwhile. In this case, Alice will probably decide to go to the meeting.

6 Make the decision and carry it out.

During the process it is often helpful to consult other people whose judgement you trust. Often just 'thinking aloud' can clarify matters. Sometimes it is important to stand back and review the sources of your value system. Usually in everyday life, time does not allow for this, but taking time out from time to time to review how you make decisions can very valuable.

In respect of fast-developing issues a further step is necessary. Decisions should be kept under review in the light of new information. A good medical example of this is that in the early days of the AIDS epidemic a patient with AIDS-related pneumonia was not treated in an intensive care unit because the prognosis for AIDS was not much more than a year. Nowadays, when drug treatment has greatly improved the prognosis, intensive care treatment is entirely appropriate under such circumstances.

Ethical decision-making can at times be very difficult. It requires careful, objective thinking if we are to make the best or the most virtuous decision. Often we have to be content not with the ideal solution but to the best one we can arrive at under the circumstances.

Summary

- Ethics is about deciding what we ought and ought not to do.

- From the earliest times, people have struggled to find the best way of making ethical decisions.

- The ancient Greeks, Christians, adherents of other religions (and of none), philosophers and scientists have all contributed to the way in which we make ethical decisions in the 21st century.

- There are three main approaches to ethics – deontological, consequentialist and virtue ethics. Right-based ethics is sometimes regarded as a fourth approach.

- Bioethics deals with the complex issues arising from the rapid developments in biomedical science in the last 50 years. Its scope also extends to environmental and global issues.

- It is important to have a disciplined framework within which to make ethical decisions.

3 Humans and the natural world

We live on a planet that has a more or less infinite capacity to surprise. What reasoning person could possibly want it any other way?

From *A Short History of Nearly Everything*, Bill Bryson (2003)

But the unintended impacts of human actions are now creating problems like global warming and the extinction of multitudes of species, problems which raise profound issues about how we should live our lives and organize our societies, and which present challenges never encountered by previous generations

From *Environmental Ethics*, Robin Attfield (2003)

3.1 Introduction

As we noted in Chapter 2, the first recorded use in print of the word *bioethics* occurred in 1971 in a book entitled *Bioethics – a Bridge to the Future* by Van Rensselaer Potter. In this book, he argued that we needed a new ethical framework, a framework that took into account the environment, and particularly the biosphere, in order to reverse the increasing environmental damage caused by human activity. In some ways he was saying nothing new. Leopold had already expressed concern about the creeping loss of wilderness in his 1949 book *A Sand County Almanac*. Rachel Carson's seminal 1962 publication, *Silent Spring*, was a call to arms for campaigning against the widespread over-use of toxic chemicals which was damaging the biosphere at all levels of complexity, and in 1967 the historian Lynn White had published a blistering critique of western attitudes to the natural world.

However, in parallel with a developing environmental concern, the concept of bioethics had been evolving in a biomedical context, especially following the Nuremberg trials. In the USA, Potter's term came to be used mainly in relation to medical and later to biomedical issues. Potter, although he was himself a biomedical scientist, was quite resentful that the word he had coined

Introduction to Bioethics, by John Bryant, Linda Baggott la Velle and John Searle
Copyright © 2005 by John Wiley & Sons, Ltd.

was, in his view, high-jacked by the medical community. However, he really
need not have worried. There was already a head of steam for the development of environmental ethics and this quickly became a subject in its own
right, with an extensive literature. However, environmental ethics clearly
overlaps with bioethics in the sense in which the latter is now used as, for
example, when we consider the use of GM crops. Our aim here therefore is
to introduce the reader to environmental ethics, whilst providing in our
reading list at the end of the book references to more detailed treatments of
the subject. In this chapter we will present some of the major current environmental problems and, where appropriate, some of the short-term solutions. We will discuss various views of the relationship between humans and
the non-human components of the environment. We will also examine some
of the current themes in environmental thinking.

3.2 What's the problem?

Introduction

At its most simple, the problem may be described in one word – humans.
Humans, although we are very obviously part of the biosphere, have a brain
power that is far in excess of even our nearest relatives. This brain power has
allowed us to exert control over the environment in a way that is impossible
for any other living creature, and as our technological expertise has developed and evolved so that ability to control has become greater and greater.
There is a good deal of debate as whether pre-technological human societies
lived in harmony with the natural environment. Certainly very early humans
were at the mercy of nature and probably did not cause any lasting changes
to their natural environment. However, as soon as societies and cultures
evolved, humans started to control nature and to use natural resources in
more organized ways. Some commentators have suggested that these developments were, as we would now say, sustainable, in harmony with nature.
This is almost a romantic view, the idealization of the 'noble savage', and
there are some who seek a return to what they perceive as being a more
harmonious, simpler existence, embodying a native wisdom that recognizes
that humans really are just part of the biosphere. We will return to the latter
idea in Section 3.4, when we discuss the different approaches to environmental ethics. However, whilst acknowledging that pre-technological humans
lived (and still do live, in some parts of the world) much closer to nature,
and had, or have, a much greater knowledge of nature and of natural
resources (such as medicinal plants) than most people in modern developed
countries, it is held by some scholars that even pre-technological control of
nature would have left its mark on the environment. Indeed, Reg Morrison
has suggested that human activities have had irreversible effects on the envi-

ronment – that they have left an 'ecological footprint' – for at least three and half thousand years.[1] Nevertheless, in 1500 BC, that ecological footprint was very much smaller and shallower than that left by modern humanity. Any damage was very much less serious and very much less widespread than happens now. Furthermore, in the great majority of cases, the effects were very localized.

However, as human society has grown and its use of technology has developed, so has the extent of environmental damage caused by human activity. Here, we deal with population in more detail. Discussion of specific aspects of environmental damage is in Section 3.7.

Population

In 2004, the earth's human population was about 6.3 billion.[2] By 2020 it is likely to have reached 8 billion. Population is a major determinant of the effects of human society on the environment. The need for living space increases as does the need to produce food. Thus 'natural' environments are taken over for human needs. Indeed, we need to look no further than the UK to illustrate this. Here much of what we know as countryside, the 'green and pleasant land', is actually a result of agriculture, and the growth of the cities, with all the attendant expansion of infra-structure and activity, including transport and leisure, is eating into both natural environments and agricultural land. However, the effects of population on environment are not all one way, as was illustrated by the huge loss of life, now estimated at well over a quarter of a million, in the Indian Ocean tsunami of 26 December 2004. Concentrations of population along the coasts and, in many areas, the destruction of natural coastal vegetation such as mangrove swamps, which would have diffused some of the force of the tsunami, combined to expose very large numbers of people to this natural disaster. Further, as is so often the case, the brunt of the disaster was mainly borne by poorer countries, with less developed infrastructure and, relevant for this case, no early warning system for tsunamis.

In addition to pressure on land there is also pressure on resources, both living and inanimate. These include the fish in the sea, forests, metal ores and minerals, rock and stone and fossil fuels. Further, the use of some of these resources has for many years taken little note that they are finite. Over-fishing, especially but not exclusively round the coasts of Europe and North America, has led to a situation in which stocks are no longer being renewed by breeding in the natural populations. For non-living resources, such as fossil fuels,

[1] Quoted by Christopher Southgate (2002) in *Bioethics for Scientists*, eds Bryant, J. Baggott la Velle, L. and Searle, J. Wiley, Chichester, pp 39–55.
[2] I.e., 6.3 thousand million.

it should be even more obvious that stocks are not inexhaustible; we cannot after all simply reproduce the conditions of past geological eras in order to make more oil or coal. Thus, several commentators have calculated that we have already passed peak oil production and yet we are still wedded to the use of oil and its derivatives to fuel the transport that is so much a feature of modern life. These examples of fish and oil are but two of many that we could have used to illustrate the unsustainable way in which human society treats natural resources.

However, the objective of sustainable living is not very easy to achieve. If humans are to live on this planet then there will always be tension between human needs and concerns about the environment. Indeed, this tension is the nub of environmental ethics. But this raises further issues. In the previous chapter we saw that conventional ethics is concerned with the way that people treat each other and in general this is based on the value that we ascribe to other human beings. In bioethics much of our thinking is therefore based on how the new biomedical technologies affect other humans. However, as will become apparent in later chapters in the book, there are sit-uations where ascribing value or ethical status is difficult, as with very early human embryos. The question of value is also at the heart of environmental ethics: how or why do we ascribe value to the environment?

3.3 Valuing the environment

In a recent edition of the British magazine *Birds*, Graham Wynne, Chief Exec-utive of the Royal Society for the Protection of Birds, stated that '*Wild birds . . . have their* own *intrinsic value*'[3] and the article also implied that the state-ment was applicable to other elements of the biosphere. In other words envi-ronment has intrinsic value; we do not need to look for any reason based on utility or consequences for humans in order to ascribe value. Such a view is widely stated by people with an interest in environmental problems, includ-ing some environmental ethicists, and it is an underlying theme in concerns about biodiversity.

Question

Why does the environment have intrinsic value?

---

[3] *Birds*, Spring 2005, pp 2–3.

This is not an easy question to answer. Wynne made no attempt to justify his statement about the intrinsic value of birds. Thus, we might say that intrinsic value is ascribed to the environment 'because it is there'. Some people would go further: the environment as we know it represents billions of years of development and evolution, going back to the origin of the universe; its very existence is both wonderful and mysterious and its present complexity represents the current state of play in an unfinished drama. Faced with this, it is said, we can do nothing else than to ascribe intrinsic value.

However, there other reasons that for ascribing intrinsic value and these reasons often involve religion. For some of our readers it may seem strange to introduce this topic, especially if they have grown up in the highly secularized society of northern Europe, but actually religion still has great influence on ethical thinking right across the world; this may be because, as in Europe and the USA, legal and ethical frameworks have been heavily influenced by Judaeo-Christian thinking, or it may be a direct cause of personal or societal religious commitment. Thus, in some eastern religions and in the pagan and neo-pagan nature religions, nature is regarded as divine. This alone is enough to ascribe intrinsic value. In the 'Abrahamic faiths' – Judaism, Christianity and Islam – on the other hand, the universe is regarded as an entity created by God but separate from God. In this instance, the intrinsic value of the environment lies in its origins as divinely created and upheld. However, somewhat paradoxically, the outworkings of this view have not always led to respect for the environment, especially in some strands of Christian thought.

Returning to the article by Wynne, the author also stated that 'Green and pleasant places have other values too, not least for the nation's health and well being and for the economy of local communities'. Here is a different approach to valuing the environment. The living organisms and the non-living components of the environment are valuable because of what they may provide for humans. The environment has instrumental value: it is valued because of its actual or potential use in supplying resources for human living. Instrumental value may include aesthetic and amenity value as in the example quoted here but more obviously includes supply of humans' material needs.

In practice, most people who have given any thought to environmental concerns have mixed views about whether the environment has instrumental or intrinsic value. Even those who espouse most enthusiastically an intrinsic value position will usually agree that in order to live at all, humans need to make use of environmental resources. Thus we move on now to consider the relationship between humans and what is often called the natural world.

3.4 The place of humans in nature

Within the living world, humans occupy a unique position. Our immense brain power has made it possible to 'tame' many aspects of nature and to make very extensive uses of natural resources, even though from time to time events such as earthquakes remind us that there are many elements of nature that humans cannot control. This unique position of the human species has tended, in developed societies at least, to isolate humans from nature and to make us forget that we are actually part of the natural order. Exactly how we regard our relationship with the rest of the natural order affects our approach to environmental issues as is apparent from the main positions in the humans and nature debate. These are

- anthropocentrism
- biocentrism
- ecocentrism
- theocentrism.

Anthropocentrism is an attitude or approach centred on humankind; the rest of nature is regarded as being there for the good of humans. It is a view that ascribes only instrumental value to the environment and, in its most extreme form, pays little regard to the effects of human activity on the environment. This view thus manifests itself in decisions about environmental ethics as a kind of species-wide rational egoism. In respect of environmental problems, it is often assumed that they will be resolved by humankind's technical ingenuity and only in the milder forms of anthropocentrism are there ideas about living in harmony with the rest of nature.

Question

Is anthropocentrism a tenable position in the face of current environmental problems?

Biocentrism is centred on the biosphere; in respect of human activity in the world it means recognizing that we are part of the natural order, that we are one of many millions of species and that our actions should take this into account. It does not prevent humans from using environmental resources; after all, every living organism does so to a greater or lesser extent. It does however, prevent us from using environmental resources as if other living things do not matter. In terms of ethical decision-making it is mainly consequentialist: we consider how a projected action will affect other living organisms.

The 'strength' of the biocentric approach varies between different proponents. Leopold, for example, placed great value on wilderness and attributed
value not to individual species but to the whole biotic community, of which,
he said, humans should simply be 'plain members and citizens'. He developed a 'land-ethic' characterized by his classic statement '. . . a thing is right
when it tends to preserve the integrity, stability and beauty of the biotic community. It is wrong when it tends otherwise'. The value of wilderness is also
a feature of the *deep ecology* movement typified by the work of Arne Naess,
but in several ways deep ecology goes further than Leopold. Naess rejected
the 'man-in-the-environment' biocentrism because it still allows humankind
to allocate value. Instead he proposed that we should adopt *biospherical
egalitarianism*, which would insist on the *real* equality of all species. This is
therefore a very radical form of biocentrism.

Question

*Is radical biocentrism compatible with living in a technology-
dependent society in the 21st century?*

Within biocentric thinking there are also strands of thought that focus more
specifically on animals. We deal with these more fully in the next chapter
but introduce them here to complete our presentation of biocentrism. The
Princeton-based bioethicist, Peter Singer, suggests that it is not good enough
simply to consider the consequences for other living things of a particular
course of action, because in the main we still tend to put human needs first.
This is what he calls speciesism, a term that implies a discrimination against
other species, a discrimination that he rejects. An extension of this view is
that other species, and especially animals, have rights. The application of
rights theory to animals is associated mainly with the philosopher Tom
Regan. Not only should we not discriminate against animals but we should
grant them rights as if they were humans. Thus, animal rights activists talk
about those who, in the course of biomedical research, carry out experiments
on animals as rights violators. The idea of granting of a range of rights to
animals has been criticized by several moral philosophers because we can
identify no corresponding responsibilities on the part of the animals.
However, those holding an animal-rights position point out in reply that we
do not deny human rights to a baby simply because he or she has no corresponding responsibilities.

A question for radical biocentrists and animal rights advocates

You have the opportunity in a house fire to rescue either a two-year-old child or a litter of Labrador puppies. Which would be your choice and why?

Anti-specieism and animal rights are further discussed in the next chapter. Here we ask some questions. Is it really possible to grant totally equal rights or ethical status to other animals? How far down the scale of complexity amongst living organisms are the proponents of these positions prepared to go? Should we avoid discrimination against, for example, the mosquito that is a vector of malaria or the trypanosome that causes sleeping sickness? Do plants feature at all in anti-specieism? Does an individual wasp or round-worm have rights?

Returning to the more general features of biocentrism, there are those who believe that it is not a strong enough position in respect of human attitudes to the environment. We need, it is said, to consider the whole world of nature, not just the living part thereof. This then leads to a position known as ecocentrism.

Ecocentrism is an ethical system centred on ecosystems. It is an attitude that takes into account the inanimate features of nature, the rocks, soil and so on. The rationale for this view is that the living organisms – the biotic component – in ecosystems are dependent on the non-living. We must there-fore value the non-living, seeing the place of humans within the total envi-ronment. There are elements of this in the deep ecology movement that we described above. Indeed, some commentators would include Naess's views as ecocentrism because of his suggestion that whole ecosystems, such as rain forests or inanimate features of the environment such as mountains and rivers, have rights. However, in the view of many environmental ethicists granting rights to such things is not tenable. Nevertheless, Holmes Rolston, sometimes called the 'father of environmental ethics', points out that we do attach value to these features and further that it is correct to say that an ecosystem is much more than the sum of its parts.

This last point is also made by proponents of the *Gaia*[4] *hypothesis*. In essence this suggests that earth's surface with its associated biosphere acts as a complex single entity – Gaia – and that our actions within the environment should be based on this understanding. It is therefore a type of global eco-

---

[4] Gaia was actually the name given by Greeks to their earth goddess, but Lovelock, the originator of the hypothesis, states that no religious connotations are intended by the name.

centrism. Furthermore, this unified entity is said to be, within limits, self-regulating, but current trends such as global warming are regarded as exceeding that self-regulatory capacity, thus 're-setting' Gaia's boundary values.

Question

Can we sustain a human population of 6.3 billion if we adopt an ecocentric approach to the environment?

Theocentrism is a God-centred approach to the world. It comes from those religious faiths in which there is belief that God is creator and sustainer of the earth, such as Judaism, Christianity and Islam. Care for the environment is then a matter of caring for something that belongs to God. However, critics of this position, even if they respect the faith communities which express this view, believe that it is really anthropocentrism in disguise because it places or tends to place humans on a different plane from other living things. Proponents of theocentrism may argue, with some justification, that the particular attributes of humans do indeed place them on a different plane from other living things but also that this gives humans a special responsibility towards the environment through a responsibility to God. Such a responsibility will certainly not ignore possible consequences for other living organisms in the making of decisions about the environment. However, it does not always work out that way, as will become apparent in the next section.

3.5 Some thoughts on stewardship

One of the influential contributions during the emergence of environmental ethics was, as mentioned briefly earlier, a paper by the American historian Lynn White.[5] His thesis was that humankind's profligate and damaging use of the environment was a direct result of a Judaeo-Christian understanding of their relationship with nature. He cited the following verses in the book of Genesis:

> *. . . fill the Earth and subdue it: and have dominion over the fish of the sea and over the fowl of the air and over every living thing that moves upon the Earth.*

The term *dominion* in particular appears to give humankind *carte blanche* to dominate and make use of the rest of nature with regard only for the good

---

[5] White, L. (1967) The historical roots of our ecologic crisis. *Science* **155**, 1203–1207.

of humans. White therefore suggested that this promoted not a theocentric (God-centred) but a highly anthropocentric (human-centred) view of the world; added to this was his view that anthropocentrism was a total disaster for the environment. Overall, then, White believed that Christianity, at least as practised in developed western nations, had a very negative influence on our attitudes to nature, and there is good deal in the way of evidence to support the view that dominion had been exercised with very little thought for nature.

However, Jewish readers of the same verses would have taken more meaning from the word *dominion*, recognizing that the Hebrew word also embodied the concept of *stewardship*. This concept means looking after or caring for something on behalf of someone else. In this instance, the implication is that humans are stewards of the environment on behalf of God; we have therefore returned to a theocentric position. Further, Jewish law embodied a respect for the land and for animals while in the Jewish scriptures (many of which are incorporated into the Old Testament of the Bible) there are frequent reminders that the world belongs to God. Many early Christians took on board this attitude of stewardship-care for the rest of nature. For example, St Francis of Assisi adopted a very egalitarian approach to other animal species, calling them our brothers and sisters. Indeed, the monastic orders in general had a high regard for and a caring attitude towards nature and the land but, in parallel with industrial development, these attitudes seems to have been lost from the wider Christian church. However, since the 1960s there has been a growing trend to adopt and practice stewardship and it is once again a major trend within Christian thinking on the environment (but see Section 3.7.3).

This discussion about specifically religious aspects of environmental ethics has been necessary because of the importance of White's paper. However, the concept of stewardship is not just a religious one. Some secular environmental ethicists have also found stewardship a helpful approach. Thus Potter's book (see Section 3.1) that introduced the word bioethics to the world was called *Bioethics – a Bridge to the Future*. The principle here is that we owe it to future generations to maintain the environment in a good state so that they too may enjoy it and make appropriate use of it. We see the same principle when there is mention of the 'trans-generational' or 'future-related' responsibilities in environmental care. Now, some moral philosophers seem to find great difficulty in considering the moral status of people who have not yet been born but to us the concept is straightforward: if we inflict permanent damage on the environment, future generations will suffer. Thus, current humans hold the environment as stewards for each other and for future generations. However, although this is certainly a workable approach it may be criticized for valuing the environment only in instrumental terms and for being very human centred (anthropocentric) in its positioning. Nevertheless, we could incorporate the idea of stewardship into a more biocentric position:

the vastly greater brain power (with all that implies) of humankind places us in a position of stewardship in respect of present and future generations of humans and other living organisms.

Summary of Sections 3.3–3.5

- There are two main ways of valuing the environment.

    ○ Some ascribe intrinsic value (value just for what it is) to the environment.

    ○ Others ascribe instrumental value (value based on its use to humans) to the environment.

    ○ Many ascribe both types of value to the environment.

- There are several different ideas about humankind's relationship with and use of nature.

    ○ Anthropocentrism is a human-centred approach that always gives priority to the needs or aspirations of humans.

    ○ Biocentrism is an approach that takes into account the consequences for rest of the biosphere of a proposed course of action; there are weaker and stronger versions of biocentrism.

    ○ Ecocentrism takes into account the effects of actions on whole ecosystems and values both the living and non-living components of ecosystems.

    ○ Theocentrism treats the environment as belonging to God.

- Stewardship (holding the environment in trust), a concept based on Jewish scriptures, has been adopted in Christian thought on the environment and by some secular environmentalists.

3.6 Two current themes in environmental ethics

There is no doubt that the human population of the world has put and is putting huge pressure on the environment. In general this is seen in two different ways. First, there is pressure on the resources that humans need in order to live. Second, there is the damage that the activities of so many humans inflict on the environment. It is in respect of these two general problems that two ideas or themes were presented in the Rio Declarations of 1992. The first is the *precautionary principle*. This is not actually a new concept; it may be and has been applied in a number of different areas. In respect of the environment the nub is that we have a duty to avoid any action that will damage the environment. It is thus a combination of deontological (we have a duty) and consequentialist (we must ascertain whether an action may cause damage) ethical approaches. There is no risk–benefit analysis of the type described in Chapter 5. If there is a possibility of harm being done we have a duty not to proceed. The principle may be phrased in a number of different ways, with slightly different emphases. One phrasing states that we should not embark on a course of action until we are sure that it is safe. This puts the onus on the proponents of any course of action to demonstrate that it will not in any way damage the environment. Another phrasing is that we should not wait for scientific proof of damage before discontinuing or preventing an activity that may be damaging. In Chapter 5 we discuss the precautionary principle in the general context of risk. Here we simply note that it is impossible to prove that any action will not cause environmental damage. Thus, what may at first sight seem a tenable approach actually becomes a complete block to almost any new development. It should of course be noted that these comments on the precautionary principle do not in any way absolve us from taking as much care as possible to minimize the risks of any development. This is an approach that has been described as the soft version of the precautionary principle.

The second theme is that of *sustainability*. In essence it means that an activity should be conducted so that it may be repeated time and time again without accumulating environmental damage. Thus we now speak of sustainable agriculture, agriculture that has no lasting effect on the environment and that therefore can be repeated from year to year. In practice, however, sustainability is often spoken of in terms of minimizing environmental damage, but this does not actually satisfy the criteria because eventually the damage may become too great to sustain. Often the motivation given for practising sustainability is that the environment must be protected for future generations; our readers will note a similarity here with the secular version of stewardship that was discussed earlier.

There are two other points to make about sustainability. First, it would be entirely possible to act sustainably at a local level while acting unsustainably on a larger scale by ensuring that environmental effects occur only at distance. Organizations that carry out activities in less-developed countries in a way that would be impermissible back home spring to mind here. Sustainability must be a globally applicable concept. The second point we present in the form of a question.

Question

There has been much talk in the early years of the 21st century of sustainable development. Is sustainability compatible with development?

3.7 Three current issues in environmental ethics

Introduction

Earlier in the chapter we discussed the problems caused by the increasing human population: loss of natural environments to supply human needs for housing, agriculture, leisure and transport and because of the over-use of other natural resources. In addition to these, human activity has also caused *direct* damage to the natural environment. Here, we deal briefly with just three examples of this:

- terrestrial and aquatic pollution
- global climate change
- environmental degradation and loss of biodiversity.

Terrestrial and aquatic pollution

During the 1960s there was an increasing awareness, first that little attention had been paid to the wider effects of the use of certain chemicals, and second that the un-regulated disposal on land and in water of industrial by-products, including poisonous metals such as arsenic and mercury, was causing damage to plants and animals and in some cases to humans as well. Indeed, the catalogue of environmental pollution since the industrial revolution is huge and gives a clear indication of the lack of thought, over many years, in respect of use and disposal of harmful compounds. Rachel Carson, in her book *Silent Spring* (1962), was one of the first to bring to the notice of a wider public the problems caused by chemical pollution. She showed for example how particular agri-chemicals accumulated in the food-chain, leading to, amongst other things, dramatic reductions in the populations of birds. Further, we now know that the dispersal of such compounds throughout the world has been very widespread. DDT for example may be detected in places where it has never been used, including Antarctica. However, in general, attitudes to this type of environmental pollution have changed considerably in most of the major developed nations. For example, the USA, the European Union and individual countries within the EU have introduced environmental protection laws in order to prevent the deliberate discharge of harmful substances into water courses and onto land. In the UK, the effects of the banning of certain agri-chemicals are plain to see in the recovery of the populations of birds of prey such as the sparrowhawk and peregrine falcon that had declined so markedly in the 1960s. That is not to say that the laws are always obeyed: some organizations will flout the law in order to save costs. Furthermore, legislation is not worldwide: there are countries where environmental protection laws are weak or even non-existent and thus the problem continues.

---

*Dilemma

The insecticide DDT is a persistent environmental pollutant, one of the 'dirty dozen' that are now either banned or soon due to be banned by the United Nations Environmental Programme. However, DDT is very effective in control of malaria and the alternatives to its use are expensive, beyond the means of poorer less-developed countries. In countries that have attempted to discontinue its use, the incidence of malaria has risen.

* Based on a study presented by Southgate, C.C.B and Aylward, A. (2002) in *Bioethics for Scientists*, eds Bryant, J., Baggott la Velle, L. and Searle, J. Wiley, Chichester, pp 73–83.

We need to state at this point that some of the effects of discharge of certain chemicals were not immediately predictable. The chemical reactions in the upper atmosphere of CFCs, until recently used in aerosol sprays and in refrigeration coolants, fall into this category. They were not expected to affect the ozone layer, the layer in the atmosphere that filters out some of the potentially mutagenic or carcinogenic ultra-violet light. But that is exactly what they did, such that there has for some years been a hole in the ozone layer, mainly over Antarctica, a long way from the countries from which most of the CFCs have been discharged. Of course, this has led to the removal of CFCs from refrigerators (at least in the major developed countries) and from aerosols, which are now often described as 'ozone friendly.' However, such cases alert us to the need for vigilance and lend support to those who are very cautious about technological innovation.

Finally in this section, we need to note that some major incidents of pollution have occurred not because of deliberate discharge but because of accident. Major oil spillages have resulted from wreckages of tankers in several parts of the world; radioactive isotopes were spread over a very wide area after the accident at Chernobyl whilst the area around the power station was rendered uninhabitable. Of course, in all human affairs accidents do happen. If we apply the precautionary principle here we might say that the implications of an accident in the generation of electricity from nuclear energy or in the transport of large amounts of oil are so serious that these activities should not occur. Is this the right way forward?

Global climate change

Many plant species respond developmentally to increased atmospheric CO_2 concentrations by reducing the number of stomata – gas exchange pores – in their leaves. It has therefore been very interesting that comparisons, within a species, of living specimens with dried specimens of various ages show a significant and increasing reduction in the number of stomata since the industrial revolution. The implication is that atmospheric CO_2 concentrations are rising, an implication borne out by the direct measurements of CO_2 concentrations in more recent years. This is directly attributable to the release of CO_2 by the burning of fossil fuels in which the carbon had been locked up for millions of years. This is exacerbated by the loss of huge areas of natural forest, which reduces the biosphere's ability to absorb CO_2 and, if the wood is burned, releases yet more CO_2 into the atmosphere. But of course it does not end there: CO_2 is a 'greenhouse gas', trapping infra-red radiation from the sun and thus heating the earth; this is the phenomenon known as global warming.

Most environmental experts believe global warming to be the most serious environmental problem of the early 21st century. Why is it a problem? After all, there is evidence that in the carboniferous era global temperatures were higher than they are today. The problem is that the increases are taking place very fast. The trapping of the carbon dioxide that led to the formation of fossil fuels took place over millions of years, whereas its release is taking place very fast. It is difficult to predict exactly what effect this will have on global climate patterns because different computer models give different outcomes. For example, some models suggest that the North Atlantic drift ('Gulf Stream') will be disturbed by currents of cold water released by the melting of Arctic ice. If this happens, northern Europe, including the UK, will become colder, even though the earth in general is becoming warmer. However, whatever the details, it is clear that the earth's climate zones are shifting, that polar ice is beginning to melt and that sea levels are likely to rise. We need to emphasize here that processes such as the melting of polar ice will not happen rapidly; even with a rise in average temperature of a few degrees Celsius, complete melting would take several thousand years. However, this does not give an excuse for complacency because first even some melting of the ice-caps will have serious consequences, and second, as we showed in Section 3.5, future generations of humans and other living organisms do need to be considered.

Further, some commentators have suggested that the warming is accelerating as increased temperatures are causing increased metabolic rates in methane-producing micro-organisms (methane being another greenhouse gas). The speed of these changes means that some living organisms may be unable to adapt and thus will become extinct, while there will also be problems for human populations, especially those living at or near sea level.

Exercise

Suggest ways to reduce global warming.

The reader should not think this is easy. Several international conferences and many national governments have grappled with the problem. In general, most nations have agreed to take steps to reduce the use of fossil fuels and to increase the use of renewable energy sources, as for example set out in the Kyoto agreement of 1997. However, fine words do not necessarily lead to effective action. The problem is complex. Modern, developed human society relies on fossil fuels in so many ways, including transport, manufacturing, warming our homes, running air conditioning units, various leisure activities and many more. With renewable sources supplying only a fraction of our current energy usage, a rapid reduction in the use of fossil fuels would necessitate such a major reduction in human activities that society as we know it could not continue. While there may be some deep ecologists (see Section 3.4) who would, in theory at least, approve of this, it would not be a generally acceptable solution. Moreover, the problem will get worse as less developed nations become more developed. The rapid changes in much of China to a modern, technology-dependent society and the steadier development in India are both increasing significantly the use of fossil fuels. This pattern will be repeated, although not on such a scale, as other countries move from less developed to developing to developed. We introduced the term 'sustainable development' in Section 3.6, but it is examples such as these that show how difficult this is.

The final point to be made in this section concerns international collaboration. Acting according to the terms of the Kyoto agreement is a step in the right direction towards reducing global warming. However, George W. Bush, president of the USA, the most powerful nation in the world, has refused to sign the agreement; he and his advisers either deny that global warming is happening, or suggest that even if it is happening it is not serious. In energy use in general, the USA heads the world: less than five per cent of the world's population is responsible for about 25 per cent of the annual global energy use. Further, a consortium representing the US oil industry is now challenging the 2001 report on global climate change, representing the work of over 2000 scientists and presented by the Intergovernmental Panel on Climate Change. Robert May, a former chief scientific adviser to the UK government, states that 'It is reminiscent of the tobacco lobby's attempts to persuade us that smoking does not cause lung cancer'.[6]

There is one further facet of this situation that deserves attention in that it highlights the ambiguous role of religion in environmental ethics. George W. Bush is widely known as a practising conservative Christian. Many of his policies are a result of the way in which he applies his faith to particular issues. The USA in general is probably the most Christian of all the northern nations and yet, as typified by the approach to global warming and energy consumption, there seems not to be any general sense of Christian stewardship (as was discussed in Section 3.5) in respect of the world's resources. For the most part, the fierce criticisms levelled by Lynn White back in 1967 still seem relevant today in the USA.

Exercise

Consider the contrasting sets of priorities between the pressure to generate energy from renewable resources (wind power, tidal power, hydro-electric power) with the pressure to conserve wild and beautiful environments.

---

[6] *The Guardian*, 27 January 2005.

Environmental degradation and loss of biodiversity

By now it will be very clear to our readers that the activities of humans have had and are having major effects on the environment. Some of these effects may seem relatively benign, such as the transformation of much of England from forests to a much more open landscape. Other effects however are clearly not benign. Pollution can affect large tracts of land and even whole ecosystems, with attendant loss of biodiversity. Even without pollution, individual species have been driven to extinction by indiscriminate hunting. Intensive use of land in agriculture without any thought of what we now call sustainability has created dustbowls, while the exploitation of some major ecosystems has led to the losses of whole assemblages of plants and animals.

These issues are well illustrated by considering *tropical rain forest*. Different variants of this type of ecosystem occur right across the tropics and they form a significant proportion of the earth's vegetation cover. These forests are rich habitats for vertebrate and invertebrate animals, for plants and for micro-organisms and contain a large representation of the earth's total biodiversity. These are 'climax' ecosystems, the final stage in the colonization of a land area, and have taken a long time to become established. They are a major 'sink' for CO_2, which is 'fixed' by photosynthesis and converted into plant material. They are also a major resource vulnerable to exploitation by humankind.

Clearance of rain forest, primarily in order to use the wood, but also in some instances to make more land available for agriculture or for human habitation, is proceeding very rapidly. Some estimates suggest that we are losing about seven million hectares per year. This is a huge loss. The area of Wales is about 2.16 million hectares, that of the state of South Carolina is 8.3 million hectares. The disappearance of large areas of forest has up to four clear-cut effects.

- A major sink for CO_2 is lost, thereby contributing to global warming (as we have already discussed).

- In some instances the local climate may change because of the loss of a major component of the water cycle.

- Soil is degraded or even lost because, without a covering of trees, erosion and run-off occur. This may in turn limit the use for agriculture.

- There is significant loss of both above-ground and below-ground biodiversity.

The obvious and simplistic solution is to stop forest clearance. However, readers should remember the case studies presented at the beginning of this book. Rain forest may be one of the few major resources in some poor countries and thus represents a significant potential for earning much needed cash. How is this tension to be resolved? In this instance, sustainable use does seem to be the way forward. This might include, amongst other things

- *managed* local scale felling rather than clear felling

- the sustainable maintenance of the forest as a genetic resource, available, at a price, for use by commercial organizations, as in Costa Rica

- promotion of non-invasive eco-tourism, again as in Costa Rica.

Finally then we come to *biodiversity*. The major loss of biodiversity caused by destruction of the rain forest is a symptom of a more general problem: extinctions are occurring at a rate that may be as much as 1000 times faster than the *average* rate through the fossil record (there have been episodes of extinction much faster than the average). Further, this estimate includes only organisms that we know about. It does not include the several million species of animals and plants that have not yet been described, let alone the myriad micro-organisms. It is thus very likely that the current wave of extinctions will include organisms that have co-existed on this earth with humankind but that humans have never seen or described.

Question

What is so wrong about extinction? From the evidence of the fossil record there have been five major episodes of extinction. Perhaps we are seeing the start of the sixth.

The answer to the question will depend to some extent on one's view of the relationship between humans and the rest of nature. What is different about the current wave of extinctions is that it is due to environmental changes brought about human activity. Further, if we ascribe value to the environment, then that extends to biodiversity. On this basis, it is not just individual plants and animals that have value but whole species and assemblages of species – biodiversity. As we noted at the beginning of this chapter, components of the environment may be valued intrinsically, for what they are, or instrumentally, for what they can provide. In discussions on biodiversity we see both, the latter being typified by concerns that we may be losing plants that could be valuable sources of medicine. We also see a strong element of stewardship or trans-generational concern: we have a responsibility to future generations in respect of preservation of biodiversity. But acting on this will not be easy.

Question

Are there any organisms whose extinction would not be a matter of concern?

3.8 Concluding remarks

However we approach environmental ethics, it is clear that it is not easy. There are always tensions between supplying human needs and caring for the environment. Different positions within environmental ethics may place different emphases within the human–environment spectrum but the tension is still there. It is the working out of that tension in tackling the range of serious environmental problems that will be the major focus of environmental ethics in the 21st century.

4 Humans and non-human animals

'Well, we're sorry,' said Mack. '. . . You see, we're workin' for some scientists. We're tryin' to get some frogs. They're workin' on cancer and we're helpin' out getting some frogs.'
   'What do they do with the frogs?'
   'Well sir,' said Mack, 'they give cancer to the frogs and then they can study and experiment and they got it nearly licked if they can just get some frogs.'

From *Cannery Row*, John Steinbeck (1945)

The question is not 'can they reason?' nor, 'can they talk?' but, 'can they suffer?'

From *Introduction to the Principles of Morals and Legislation*,
Jeremy Bentham (1789)

4.1 Introduction

In the last 50 years, growing awareness of animal welfare issues combined with the activities of animal rights protesters has brought about significant changes in attitudes to the use of animals in the UK. For example, there are bans on fur farms and on using live animals in cosmetic testing, and now fox hunting in England and Wales has been prohibited by an Act of Parliament that came into force early in 2005. However, 50 years is a very short time in terms of human history, and through nearly all that history human beings have lived with, relied on, cherished and exploited animals. In fact it is true to say that other members of the animal kingdom are vital to us. So why have attitudes towards them changed to dramatically in recent years? How has political reaction to animal issues been so quick? How have the protesters managed to become so effective?

   In evolutionary terms, humankind, *Homo sapiens* is a very successful species. We can and often do dominate all the ecological niches we care to inhabit, and there is a fast growing number of us. However, as organisms we

Introduction to Bioethics, by John Bryant, Linda Baggott la Velle and John Searle
Copyright © 2005 by John Wiley & Sons, Ltd.

are dependent upon other organisms, and the history of our success has been founded upon our exploitation and manipulation of other species. This fact lies at the heart of several bioethical issues, various of which are considered through the chapters of this book. This chapter will consider some features of our association with other animals, especially vertebrates and particularly mammals, and the ethical problems that these relationships create.

4.2 Humankind's place in the animal kingdom

As all who have studied biology during their secondary school (high school) education will know, all living things share most of the common characteristics of life: movement, respiration, sensitivity, growth, reproduction, excretion and nutrition. So what distinguishes humans from other members of the animal kingdom? The main characteristics are probably self-consciousness and the higher cognitive powers of being able to use language, and through this to rationalize. While animals can obviously communicate with one another, it also seems clear that their level of communication has a limited vocabulary. There is some evidence that our closest relatives among the primates can understand human language. Washoe the chimpanzee and Koko the lowland gorilla for example have been successfully trained to respond to American Sign Language, and there are many reported instances of communication in other animals such as dolphins and parrots. However, while we may not be alone in being able to learn and use language, there seems little doubt that the level of sophistication reached by even the simplest of human languages far exceeds any communication system developed in any other species.

Many philosophers have claimed that humans alone have reason, and that this is an all-or-nothing state. Reasoning in this sense goes beyond the ability consciously to make decisions relevant to our survival, because other animals are clearly able to do this. We can also decide what is right or wrong, in other words moralize. This has, probably from the beginning of the history of our species, shaped the development of human communities and civilizations. The development of language has facilitated, enhanced and augmented the human ability to moralize through reasoning. In addition, although the social and cultural climate in the UK may hide the fact, a very high proportion of the world's people have a religious faith, a factor that is also relevant to this discussion and, as has been mentioned earlier in this book, to ethics in general. Indeed, the philosopher Mary Warnock has said that even in our very secular age we should not forget the influence that religion has had on ethical thinking.[1]

[1] Warnock, M. (1998) *An Intelligent Person's Guide to Ethics*. Duckworth, London.

Most religious people believe that humans have a soul or spirit or spiritual dimension, which may be described variously as the divine Self in Hinduism, as the product of conditions and causes in Buddhism, as the core of the individual person, influencing his or her choices and deeds, in the Abrahamic faiths (Judaism, Christianity and Islam). Addressing questions of the nature of spirituality and its relationship to mind and consciousness are well beyond the scope of this chapter, but it remains among the most profound and intriguing matters facing philosophers, scientists and theologians today. The increasing interest among neuroscientists about consciousness has given momentum to the debate, but whether consciousness is a uniquely human characteristic is at present an unanswerable question.

Because of these beliefs about the human condition, throughout history most cultures have come to regard people as being of greater intrinsic value than other animals, and have therefore justified their use of them for food, experiments, fashion and fun. Further, as will become apparent, attitudes to animals in western society have until recently been very much influenced by a particular Judaeo-Christian understanding of the role of humans (see also Chapter 3).

4.3 Human use of animals

Historic and present day perspectives

Across the world and within all cultures people use animals as pets, beasts of burden, subjects in experiments, objects of study, of reverence, for financial gain and as the basis for subsistence. Thus humans have taken control of the lives of animals. How has belief in this right to commodify animals, even to the extent of determining the existence or disappearance of an entire species, come about? Again, the answer to this has its roots in religious thinking, which although it again may not be immediately apparent has greatly influenced attitudes towards animals for many centuries. As we noted in the previous chapter, much weight has been placed in Jewish and Christian thought on the following:

> Be fruitful and multiply, and fill the Earth and subdue it: and have dominion over the fish of the sea and over the fowl of the air and over every living thing that moves upon the Earth.[2]

The significant word here is *dominion*, and the idea that it is right that humans should dominate all other species probably springs from this. However, as was discussed in Chapter 3, the term *dominion* has been misinterpreted over the years in that the sense of *stewardship* that is inherent in the original Hebrew

---

[2] *Genesis* Chapter 1 v 28.

word has been ignored. It has however been re-discovered in the second half of the 20th century and on into the 21st with concomitant effects on thinking about relationships with the rest of the natural world.

Many belief systems have also been influenced by Greek thinking and in particular that of the philosopher Aristotle (384–322 BC). He first proposed the idea that humans alone are rational, and suggested a natural hierarchy in which inanimate objects such as stones were below plants, which are alive; next came animals, which Aristotle deemed to be sentient. Above animals came humans, who are rational, and at the top was the state of perfect reason, occupied by a divine mind. Aristotle saw rationality as a godly virtue, so humans alone among organisms in his view have a divine element within them. In many faiths this is seen to be the soul and it is this that makes us humans able to act morally and have an understanding of concepts such as good and bad, just and unjust.

In early Christian thinking, St Augustine (AD 354–430) was very influential. He taught that the commandment *Thou shalt not kill*[3] does not apply to animals, because animals cannot reason and are different from humans because of this lack of rationality. Augustine thus believed that God subjected animals for the good of humans and that it is also right for people to keep animals alive for our own uses.

The French philosopher, Rene Descartes (1596–1650), put forward a major interpretation of the relationship between humans and other animals, although many disagreed with him. His theory was based on the ideas that the mind and soul are one and the same thing and that possession of a mind or soul is an all-or-nothing matter, uniquely human. He thought that animals were akin to machines – operating without consciousness, and that it was therefore not morally wrong to exploit them. However, most philosophers of the time agreed that animals could suffer and that inflicting suffering on them was wrong.

Emmanuel Kant (1724–1804) believed in the idea of personhood as the quality that makes a being valuable and thus morally important. He thought that humans have no *direct* duties to animals because animals are not self-conscious, so they cannot make judgments. The work of Carolus Linnaeus (1707–1778) may have informed Kant's views. Linnaeus, sometimes known as the father of taxonomy, classified animals in a hierarchy of perfection, at the top of which was man. Although his ideas predated any notion of evolution, Kant held the view that animals are a means to an end, and that end is man so therefore animals may be man's instruments. However, he did qualify this by stating that it is sometimes wrong to hurt animals. Kant believed that the way we treat an animal might affect or determine how we treat other humans. This has become known as the "indirect duty" view of human/animal association. This idea suggests that the wanton inflicting of

---

[3] *Exodus* Chapter 20 v 13.

suffering actually harms the perpetrator. Put another way, we should be kind to animals not because of our duty to them directly, but because it is good practice for being kind to humans, who *can* judge us. Kant reasoned that we have no duties directly towards animals, not even those of compassion or sympathy, but we do have a direct moral obligation to other humans for compassion, and one effect of this will be to improve our society.

The English lawyer and philosopher Jeremy Bentham (1748–1832) changed the views of many people's attitudes towards animals. Rather than regarding them as inferior to human beings because of their inability to reason as both Descartes and Kant had done, Bentham stated that 'The question is not "can they reason?" nor, "can they talk?" but, "can they suffer?" '. He said that because animals were capable of pleasure and pain, their happiness was morally relevant. Arguing that considerable evidence showed that animals might suffer, Bentham reasoned that in a humane society they should be given protection. This is a utilitarian view (see Chapter 2), and the change in thinking it represents has underpinned all contemporary animal legislation in the UK. In 1822 the first Act of Parliament to outlaw cruelty to animals was brought in by Richard Martin, who went on two years later to found the Society for the Protection of Animals, which subsequently became the Royal Society for the Prevention of Cruelty to Animals.

Through the 19th century people were thinking more about the welfare of animals. The English physician and physiologist Marshall Hall, who as a result of experiments on animals was the first to describe reflex action, proposed a code of ethics for experiments in 1831. Queen Victoria was apparently opposed to animal research, as she asked Joseph Lister to address a Royal Commission Enquiry into vivisection, asking him to 'make some public statement in condemnation of these horrible practices'. This request presented a particularly difficult dilemma for Lister. As a devout Quaker he condemned practices such as killing animals for pleasure, slavery and the cruel treatment of offenders and the mentally ill. However, he recognized the need for animal experimentation, not only in his own research but to accomplish other advances in medicine and in scientific knowledge. He testified before the Royal Commission in 1875, emphasizing the fact that restricting research in animals would prevent people from making discoveries that would benefit humanity. In the following year the Cruelty to Animals Act was introduced, and among other things this law required that animal researchers must apply annually for a licence to practise, and that any procedures that subjected animals to pain must have special permission.

By the end of the 19th century some writers, such as Lewis Gompertz,[4] himself an early SPCA member, and Henry S. Sale,[5] defended the rights of animals, but it was not until the second half of the 20th century that the

[4] *Moral Inquiries on the Situation of Man and Of Brutes* (1824) Mellen, Lewiston, NY.
[5] *Animals' Rights* (1892) Society for Animal Rights, Battimore, MD.

strong defence of animals' interests became an increasingly important issue in the public debate. In 1975 the Australian ethicist and philosopher Peter Singer (now at Princeton, USA) put forward the view that animal and human interests are comparable in moral terms, arguing that since a difference of species entails no moral distinction between sentient beings, it is wrong to mistreat non-human animals; it therefore follows that animal experimentation and the eating of animal flesh are morally indefensible.[6] Singer argues that the capacity for suffering is the vital characteristic that entitles a being to equal consideration. If we define suffering as the susceptibility to pain or awareness of being in pain or about to be in pain, there is little doubt that most vertebrates can suffer. However, the extent to which they are aware is questionable, but there is good evidence to suggest that the great apes have a high degree of self-awareness. This raises the question of the extent to which sentience contributes to consciousness, and thus also raises the question of the extent to which consciousness is a purely human characteristic. Emanuel Kant thought that personhood was the defining characteristic of humanity, and Singer believes that the distinction of personhood should be decided on the basis of whether or not a being is self-conscious. This immediately raises the problem of whether this state is attributable to all humans but to no other animals.

Speciesism

Some members of humanity, i.e. babies and little children and people with certain forms of autism or Alzheimer's disease or other cognitive disorders, do not have the rational, self-reflective capacities associated with personhood, but does that make them less human? Peter Singer argues that logically in this respect the intellectually impaired, the disabled, infants and embryos are akin to non-human animals. He has controversially used this argument as the basis for equating the justification of animal experiments with them. A central tenet of his opinions is that if we say it is alright to use animals for our own ends in ways which cause them to suffer just because they belong to another species – because they are *only animals* – then we are showing a form of prejudice akin to racism or sexism. Thus, as we saw in the previous chapter, the word speciesism has been coined, meaning human intolerance or discrimination on the basis of species, especially as manifested by cruelty to or exploitation of animals. To those who believe in it, speciesism underlies all our uses of animals that cause them harm.

The contemporary American philosophers Tom Regan[7] and Evelyn Pluhar[8] oppose claims that human beings alone are rational and therefore entitled to

[6] *Animal Liberation* (1975) Ecco, New York.
[7] *The Case for Animal Rights* (1985) University of California Press, Berkeley, CA.
[8] *Beyond Prejudice: the Moral Significance of Human and Nonhuman Animals.* (1995) Duke University Press, Durham, NC.

superior moral status. Regan defends the inherent value of all living individuals and also decries speciesism, which pretends to separate human from
nonhuman animals. Independent of any benefits humans might derive from
exploiting animals, Regan argues that, on a philosophical level, there is no
sustainable defence for separating human and non-human animals as beings
of absolute, as opposed to instrumental, value. He offers a detailed analysis
and critique of Peter Singer's philosophy, and then puts forward an alternative understanding of humanity's moral obligations to animals. Regan develops the idea of animal rights, arguing that animals possess morally important
characteristics, and those that we use for food, experiments, sport and fashion
all have inherent value, which is equal to our own. Animals have an equal
right to be treated with respect, not to be used as mere resources. Regan
argues that this right is violated by our current practices and goes on to call
for a total abolition of the use of animals in science, agriculture, and sport.
Regan rejects the indirect duty view of Kant's and Singer's utilitarian ideas
because he believes that a good end does not justify evil means. Some people
however claim that equating speciesism with racism and sexism is basically
wrong because there is a moral distinction between humans and animals, for
example that human pain has more moral importance than animal pain.

Vivisection

We need to start the discussion of this contentious issue by defining our terms.
Vivisection is the performance of surgical experiments on living animals in a
laboratory for the advancement of (especially medical) knowledge.[9] An experiment is a scientific investigation that has an uncertain outcome. With respect
to vertebrates, (and interestingly one species of octopus), in the UK this type
of procedure can only be done if it is licensed under the Animals (Scientific
Procedures) Act of 1986. Many invertebrates such as fruit flies and worms
are also used in research, but are not protected under British law. The Act
covers any procedure that is likely to bring harm or suffering to an animal,
and around 2.7 million such procedures per year fall into this category,
although there is evidence that this figure is in steady decline (see Figure 4.1).

Animal experiments cover all types of investigation from the testing of the
toxicity of a new oven cleaner through to verifying the effectiveness of a
newly discovered vaccine. These activities take place in a variety of venues,
including pharmaceutical laboratories, public research institutes and university medical schools. They are frequently the focus of protest activities and
in recent years well organized campaigns have, sometimes with increasingly
violent actions, succeeded in intimidating employees and suppliers of some

[9] *Oxford English Dictionary.*

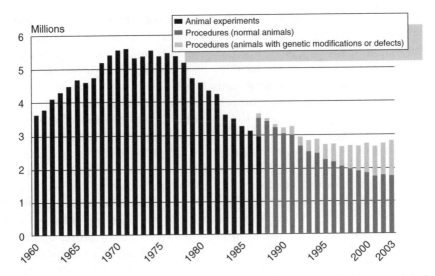

Figure 4.1   Total numbers of animal experiments would appear to be in general decline, but those involving transgenic animals are gradually increasing (data and graph from the Research Defence Society: www.rds-online.org.uk)

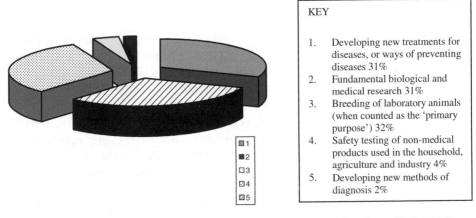

KEY

1.   Developing new treatments for diseases, or ways of preventing diseases 31%
2.   Fundamental biological and medical research 31%
3.   Breeding of laboratory animals (when counted as the 'primary purpose') 32%
4.   Safety testing of non-medical products used in the household, agriculture and industry 4%
5.   Developing new methods of diagnosis 2%

Figure 4.2   Areas of research involving animal experiments (data from RDS, 2003)

pharmaceutical laboratories and animal breeding establishments to the extent that research in the UK may be under serious threat.[10]

As the breakdown of numbers in Figure 4.2 shows, the vast overwhelming majority of animal procedures are concerned in one way or another with

[10] *UK animal protests scare off drug firm suppliers.* Reuters article 19 Jan 2005.

research into human and animal disease – its causes and the development of treatments for it. Product testing is a very small, but to some people very important, element. Health and safety legislation requires that the wide range of everyday chemicals used in household, medical, agricultural or manufactured products must be tested to ensure that they are safe to use and to handle. This process is essential to avoid, for example, causing cancer or birth defects in both humans and animals. Testing of these commodities accounts for nearly half of the total number of tests. Environmental protection is also important in this regard, and accounts for about a third of all testing done. Food and food additive safety involves about 7% of total testing. There has been no testing of cosmetics or toiletries since 1998.

But what animals are actually involved in these procedures? Are they of equal ethical importance? Again figures from the Research Defence Society (RDS), which exists to promote the understanding of the use of animals in biomedical research, indicate that the overwhelming majority are rodents, specially bred for the purpose (Figure 4.3).

The difference between these species of animals may have ethical importance in this argument. Although Darwin said 'the difference in mind between man and the higher animals, great as it is, certainly is one of degree and not kind', surely the complexity of this relationship has its foundations in more than phylogenetic taxonomy. Carrolus Linnaeus's ordering of organic forms in the volumes that comprise the *Systema Naturae* (1735) was developed in a large measure to categorize the types of organisms, animals included. The Linnean system was meant to showcase the 'Creator's handiwork', how each set of animal types led to higher and higher types (from slugs to man) in a ladder-like rise to perfection. Pre-Darwinian scholars defined species in a way that was closely linked to their theological views on the origin of the uni-

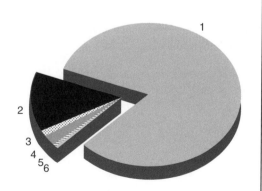

KEY

1. **85%** Rats, mice and other rodents. All specially bred laboratory species
2. **11%** Fish, amphibians, reptiles and birds (including many fertilized hen's eggs)
3. **1%** Small mammals other than rodents, mostly rabbits and ferrets
4. **2.7%** Sheep, cows, pigs and other large mammals
5. **0.3%** Dogs and cats. Specially bred for research. No strays or unwanted pets can be used
6. **0.17%** Monkeys, such as marmosets and macaques. Chimpanzees, orang-utans and gorillas have not been used in the UK for over 20 years and their use is now banned

Figure 4.3 Numbers of different animal species used in regulated scientific procedures in the UK in 2003

verse. We are ourselves mammals, and so it is not unreasonable that we may feel a closer affinity to other mammals than we would for example to amphibians or fish. Among the mammals, certain species have become historically and culturally particularly close as we have tamed and domesticated them, so cats and dogs to some people have a greater importance than for example, rats. However, as we have seen, there are those who profoundly disagree with this. The next section of this chapter will deal with the ethical arguments for and against using animals in biomedical research, and will discuss these areas of dissent and also suggest points of agreement, which may provide the common ground necessary for advancement of the debate.

4.4 The ethics of animal research

Those in favour of using animals in research into human health and welfare – let us call them the biomedical lobby – cite a number of arguments, each of which have been refuted by those opposed to it – let us call them the animal rights lobby. Table 4.1 summarizes these arguments.

In spite of the heat generated by debate, which often spills over into violence, between these two extremes of view is a surprising amount of common

Table 4.1   The arguments for and against the use of animals in biomedical science

Biomedical lobby	Animal rights lobby
Human Life is intrinsically more morally valuable than animal life: we are more important than them	All sentient animals have equal moral worth: their lives are as valuable as ours
No major medical advance possible without animal experiments	Animal experiments make little or no difference to human life expectancy or disease rate
There are no scientifically valid alternatives	Non-animal alternatives such as *in vitro* studies, epidemiological studies and computer models have validity
New medical and scientific practitioners cannot be trained without using animal experiments to mimic surgical or physiological processes in humans	Students are often desensitized by the educational process
The existing legislation provides adequate protection against the undue exploitation of animals	There is clear evidence of cruelty
Research ethics committees of funding bodies are rigorous in their consideration of animal welfare when deciding where to deploy their limited monies	Much research is trivial

ground. This has been listed by David DeGrazia,[11] who suggests that areas on which the biomedical and animal protection communities can agree may form a useful platform for future dialogue. The first principle is that ethical issues are raised by the use of animals in research, and few would doubt that human health as the main goal of such research is of ethical importance, as is the use of government funding of such research. However, important though human health is, its ends are not justified by all means. For instance, there is no doubting the instant reaction of repugnance to the notion of human experiments.

The next point of agreement is that sentient animals deserve moral protection, and that not doing so offends many people's sensibilities. This is recognized by most governments in countries where biomedical research involves the use of animals, and enshrined in these countries' legislations. Further to this is that animals' quality of life ('experiential well-being') itself deserves protection. Everyone recognizes that some social animals suffer considerable deprivation if kept in isolation. Apes, monkeys and canines for example develop social structures such as hierarchies and alliances and maintain long-term relationships that are very important to them. Some species deserve particularly strong protection. Examples would include endangered species and individual higher primates that are no longer useful for research, such as older laboratory chimpanzees.

The biomedical community often cites as the guiding principles of research the '3Rs' argument, i.e. refinement – to make sure animals suffer as little as possible, reduction – to minimize the number of animals used – and replacement – to replace animal procedures with non-animal techniques wherever possible. The animal protection lobby would not disagree with this.

Only the most vehement animal rights proponent would disagree with the idea that there are significant moral differences between humans and animals. This is bound up with the notion of autonomy, which in effect is only enjoyed by competent adult humans. Children and animals, though sentient, are not covered by this principle; for example, it is sometimes appropriate to limit their freedom of action by preventing a child from running into the road, or limiting an obese pet's diet. Common ground for most in this argument can be extended to the idea that the moral presumption in favour of life is stronger in humans than at least most animals, in other words, few would disagree that it is worse to kill people than to kill some animals which have moral status. It is interesting to note that DeGrazia uses the relative pronoun *who* instead of *which* in this context, attributing personhood to animals.

Finally, DeGrazia recognizes that both communities would agree that some animal research is justified, particularly that where no harm is done to the animal, such as observations of their natural behaviour in their normal habitat.

---

[11] In *The Animal Ethics Reader*, eds Armstrong, S.J. and Botzler, R.G. (2003) Routledge, London.

Question

Can any of the following 'direct actions' on the part of animal rights supporters be justified:

- *Threatening the lives of scientists and their families*

- *Intimidating employees of contract research companies*

- *Releasing laboratory animals*

- *Targeting the financial basis of a pharmaceutical company*

- *Persuading other companies to stop supplying them*

- *Demonstrating at their gates – you might like to speculate as to what sort of demonstration behaviour is un/acceptable*

In spite of all these areas of agreement, however, many animals are still used in experiments and animal rights supporters continue to besiege pharmaceutical laboratories. Considerable sums of money have been invested in the search for alternative methods, and there has been progress in some areas. Fewer animals are now used (see Figure 4.2); for example, the much-reviled LD50 test[12] has now been virtually abandoned, and although every new drug for human use has to be tested on two species the practice of using lower vertebrates and some invertebrates and even in some instances bacteria is gaining ground. The use of statistics and computer modelling as well as cell culture and newer technologies such as magnetic resonance imaging (MRI) to probe the human body non-invasively are increasingly used. At present, however, most biomedical scientists, while they accept that minimizing the number and degree of suffering of laboratory animals is very important, cannot actually see an end to their use.

---

[12] LD50: up to 200 *rodents* would be given increasing doses of the test substance to determine the dose that would kill exactly half of them.

Question

Comment on the following data for an LD50 test using dioxin:

- *Guinea pig – 1 microgram/kilogram*

- *Hamster – 5000 micrograms/kilogram*

- *Female rat – 45 micrograms/kilogram*

- *Male rat – 22 micrograms/kilogram*

4.5 Animals in sport, companionship, leisure and fashion

These are all very important aspects of human association with animals, but none is free of ethical implication. In the UK Parliament recently the debate about foxhunting has come to a head with the passage of the Hunting Bill (2004), which has outlawed, in England and Wales, the pursuit of wild mammals with dogs. This represents the culmination of many years of campaigning by anti-hunting protesters, who see the sport as cruel and divisive of society. Hunting people on the other hand defend the right to choose how they pursue their lives in the countryside and have themselves protested in their hundreds of thousands about what they see as an infringement of their liberty and in some cases livelihood. The pro-hunting people would argue that they have a genuine interest in the welfare of animals and that hunting is a natural way to maintain a healthy population, but its opponents believe that other methods of control are more humane. 'Country sports' supporters are worried that shooting and fishing may be next to be banned. All these traditional pastimes have characterized the UK countryside for hundreds of years, but society's attitudes have changed. This issue has now become highly politicized, and seems set to be contentious for some time to come.

The Grand National, held every spring at Aintree, near Liverpool, is a famously arduous horse race: a steeplechase, over four miles (ca. 6.5 km) long, with over 30 large fences to jump. Each year it attracts protesters claiming that it is cruel. Racehorses and other competition horses such as show jumpers and eventers are highly bred and rigorously trained to peak fitness. They can very easily suffer injuries in their sports, and often these injuries are difficult or impossible to treat. Sometimes it is economical to try to save a horse for breeding, but generally if the injury is severe owners will cut their losses and have the animal destroyed (i.e. shot). Animal protection groups see this as a form of cruelty that should be stopped and increasingly organize protests at race and other equestrian meetings. The horse, throughout history, has held a particularly important place in people's lives, and since the industrial revolution sport has taken the place of work for the horse.

Question

Is it ethical to breed and train horses for such sports as racing and eventing?

Companion animals, such as dogs, cats and caged birds, come into a similar moral category as the horse, although perhaps the physical strains that people put upon them many not always be so great. Ethical questions have been raised about certain breeding programmes to produce breed characteristics for people's aesthetic satisfaction. For example, the English bulldog has become so refined by selective breeding that it has difficulty in breathing and giving birth naturally.

For some people, companion animals such as dogs and cats have the status of children or friends. As such, if they are kept in the home, owners have the moral obligation to promote their welfare. Occasionally, this objective strays into areas where the animal may be harmed, for example when it is excessively pampered. Some people believe that pampering pets is a form of cruelty. In some cases, where their health is affected, such as happens when an animal is overfed, this may well be true, but subjecting them to for instance beauty treatments is more difficult because many pets may enjoy this sort of attention. However, there is no doubt that this constitutes a subversion of their natural behaviour.

Question

Is it always wrong to pamper a pet?

Direct action with its attendant publicity has ensured a high profile for opponents of the fur trade. Long regarded as a luxury item of clothing, fur has become less fashionable since people's awareness of animal welfare and conservation issues has been raised. Nevertheless, although the big cats such as leopard and jaguar are protected, there is still a demand for spotted cat skins for the fashion trade, and in the USA thousands of wild lynx and bobcat are trapped for their fur each year. Fur farms are now largely outlawed in the UK, although wild mink, themselves escapees from fur farms, are regarded as a nuisance, and trapped and hunted.

4.6 Animals for food

If you are a meat-eater and live for at least 70 years, it has been estimated that in your lifetime you will eat 30 sheep, 30 pigs, 600 chickens and a small herd of 5 cows. In developed countries, the rise of the supermarkets and particularly their purchasing power means that relative to income food has never been cheaper. This is one of the factors contributing to the increasing rates of obesity and related illnesses such as heart disease and non-insulin-dependent diabetes. There is growing concern about these issues, some of which are discussed in other chapters of this book. Here, the ethical problems stem from the conditions by which the meat, poultry and fish arrive in such plentiful, affordable and sanitized arrays on the supermarket shelves.

CIWF[13] is an international organization that seeks to achieve the global abolition of factory farming and the adoption of agricultural systems that meet the welfare needs of farm animals, in the belief that this will also benefit humanity and the environment. Their aim is to achieve the ending of not only factory farming systems but also other practices, technologies and trades that impose suffering on farmed animals. By hard-hitting campaigning, public education and vigorous political lobbying they have brought about reviews and in many case outright bans on such animal welfare issues as prolonged animal transportation for slaughter or further fattening, the practice of keeping sows tethered in gestation pens, battery cages for laying hens, veal crates and fur farms.

Keeping the balance of interests between animals, the environment, the farmer and the retail chain is among the most complex of current political challenges, and at the various levels of well-being, autonomy (choice) and justice (fairness) for each of their constituencies is fraught with ethical dilemmas. For farmers, their well-being will depend upon whether they have a satisfactory income and whether their farms thrive; their autonomy will be through their managerial freedom and their exercise of justice through fair trade rules by which, it is hoped, they sell their produce. For the animals, their well-being will depend upon the care given them, their autonomy through the degree to which they can engage in their natural behaviour, while justice for animals resides in respect for their intrinsic value (see Chapter 3). For the farmers and the animals, the forces exerted by the well-being,

---

[13] CIWF: Compassion in World Farming.

autonomy and justice for the consumers, *via* the retail trade, cause the ethical problems. For example, bacon is cheaper for consumers (their well-being and autonomy) if the breeding sow is kept tethered in a stall to prevent her from rolling onto the piglets and crushing them. However this compromises the well-being and autonomy of the sow and the practice of justice towards her. Nevertheless, it is possible that the consumers, through exercise of their own free choice, may actually opt to pay more for bacon in order to protect the welfare of the sow. This is therefore another indication of the complexities and possible clashes of principle that occur in much ethical decision making.

Question

Briefly describe the ethical issues raised by the interests of (a) a dairy farmer and (b) a dairy cow in terms of well-being, autonomy and justice for each of them. Would these issues be different in a developing country?

4.7 Conclusion

This chapter has considered the historical associations of humans and animals and the moral and ethical issues that arise from their places in our lives. Perhaps we can put the animals we have discussed into three categories:

- wild animals
- domesticated animals
- companion animals.

Although few examples of ethical matters concerning wild animals have been given here, issues raised by their conservation are discussed in Chapter 3. It is the impact on the lives of those animals that we have tamed and bred for our own ends, and to whom we may have a moral duty of care, that has been the subject of this chapter. The tension between animal welfare and animal rights has been apparent throughout: animals can be of benefit to humans (welfare), or animals should not be used by humans (rights); animals should be treated as humanely as possible and not caused unnecessary pain (welfare), or no matter how much they may gain, humans should not inflict death or pain on any animal (rights); we should improve, but not radically change our view of animals (welfare), or we should change our treatment of animals, even if it means a radical shift in our attitudes towards them and the way we use them (rights). As we have seen, there is a range of ethical positions on the many-faceted range of human and animal associations, which may make it very difficult to decide on one's own moral position. In the final analysis, it is a personal matter, but as we learn more detail of the genetics, physiology, behaviour and ecology of those other animals with whom we share this world it will become an increasingly challenging task.

5 Biotechnology and bioethics

'Well,' said Pooh, 'if I plant a honeycomb outside my house, then it will grow into a beehive.'

Piglet wasn't sure about this.

'Or a piece of a honeycomb,' said Pooh, 'so as not to waste too much. Only then I might only get a piece of a beehive, and it might be the wrong piece, where bees were buzzing and not hunnying. Bother.'

Piglet agreed that that would be rather bothering.

'Besides, Pooh, it's a very difficult thing, planting, unless you know how to do it,' he said.

From *The House at Pooh Corner*, A.A. Milne (1928)

5.1 Introduction

The term *biotechnology* as used here is a relative newcomer to the vocabulary of science, entering into regular use in the late 1970s. But what exactly is it? One of us, during a holiday in Crete in the mid-1980s, noted that a closely related word, *biotechnica* (as trans-literated from the Greek) actually referred to agricultural machinery. However, biotechnology does not mean tractors and combine harvesters. In a UK government report published in the early 1980s the definition included 'the use of biological systems in the provision of goods and services'. Such a definition covers a huge range of activities and could, for example, include several traditional industries such as brewing, wine-making and baking, all of which are dependent on the activities of yeast cells. However, these industries were not really part of what most people regarded as biotechnology. The term had come into use following the invention in the early 1970s of genetic modification techniques with the attendant expectation that these techniques would be a source of income-generating activity in the biology-related industries, especially pharmaceuticals. Thus in the minds of many people biotechnology has been equated with genetic modification, but actually it encompasses a much wider range of activities, finding applications as diverse as pregnancy testing kits and the remediation of contaminated land. However,

Introduction to Bioethics, by John Bryant, Linda Baggott la Velle and John Searle
Copyright © 2005 by John Wiley & Sons, Ltd.

the focus on *genetic modification* provides an opportunity to discuss a number of general ethical issues. These will form the bulk of this chapter, but we will also cover two up- and-coming developments, namely *nano-technology* (for example the minituarization of drug delivery systems) and *cybernetics* (the direct interfacing of biological systems with computer technology).

5.2 General ethical issues related to genetic modification

Introduction

When the first papers on genetic modification were published they raised a huge flurry of interest across the bioscience and biomedical communities. Some of this was related to the research potential of these new techniques. Some was related to its commercial potential (as discussed in the next chapter). However, some of the interest was certainly ethical. Indeed, in the UK in the late 1970s it was not uncommon for students to be set essays along the lines of '*Discuss the ethics of genetic engineering*',[1] even though most biological scientists were unaccustomed to talking about ethics and many would have been out of their depth in discussions of ethical theory or moral philosophy. Nevertheless, it is from this ethical interest that we can trace one of the strands of bioethics as it is now practised (see Figure 5.1).

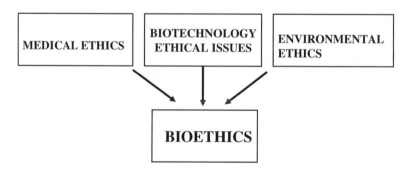

Figure 5.1   Evolution of bioethics

---

[1] There has been an evolution of the terminology over the years, from *genetic engineering* to *genetic manipulation* to *genetic modification*, possibly in attempts to lessen any negative connotations. It has been suggested that the scientists should have stuck with *recombinant DNA technology*, which is the general technical term for this range of techniques.

Within the general ethical interest in genetic modification we can detect three main components, namely

- ethical analysis of genetic modification itself
- risks associated with genetic modification
- possible misuses of genetic modification.

We now deal with these three in turn.

Ethical analysis of genetic modification

Question

Are there any intrinsic ethical objections to genetic modification?

This question was for some the central point of the debate. Does genetic modification take humankind's ability to alter nature a step too far, thus transgressing boundaries that should never be crossed? Certainly some believed and some continue to believe this to be the case, and although many of the arguments are consequentialist in nature (for example, related to risk, as discussed in the next section) some of those who hold this position have intrinsic objections to the whole idea of moving genes.

There are several possible reasons for holding this view, based on different ideas about nature. For example, some very conservative religious views embody the idea of species as being fixed entities. Further, even amongst those who not hold that species are fixed entities, there is the view that an organism's genes are part of its essential nature, its *telos*, and that genetic modification distorts that essential nature. Others have a view of nature that regards the concept of a gene as a moveable entity as being far too reductionist; in this view, any gene is part of a complex web of life and moving it into another organism will disrupt that web and may thus disturb 'the balance of nature'. Finally there are those who simply consider such activities to be 'off limits' for humankind, a view generally based on particular views of the relationships between humans and the natural world. Although many of our readers may not have encountered these intrinsic objections to genetic modification, they are certainly around and indeed are sometimes vigorously presented in discussions of GM technology. This raises some interesting questions about the relationship between private morality and public policy.

> ## Questions
>
> *To what extent should the scruples of a minority be taken into account in respect of the application of GM technology? Should there be for any product whose manufacture involves GM techniques an equivalent that does not? If so, would it matter if the non-GM version were less effective or less safe than the GM version (as, for example, for a range of vaccines)?*

Genetic modification and risk

Although some intrinsic objections to GM have been expressed, albeit by a small minority, much more widespread have been concerns about risk. Would the introduction of a gene into the genome of another organism have unforeseen effects? Was there the possibility of escape, for example of new forms of micro-organisms that posed dangers to humans or to the environment? What about the possible consequences of introducing genes into crops that made them more vigorous and able to compete in the wild? We could go on to construct a very long list of possible risks of GM technology, risks that will be different between different recipient organisms and according to the gene that is transferred. There was and still is the risk of specific misuse, a topic that we discuss in the next section. However, the scientific community was in fact well aware of these risks and in respect of GM techniques scientists took the unprecedented step of calling a temporary halt to research and development. An international (but mainly American) conference was held at Asolimar, CA, in February 1975, to debate the safety issues and the best ways of dealing with them. How one considers the outcome of the conference depends on one's views of genetic technology. Those who had strong objections believed that the conference was a device to make it appear that the scientists were concerned about possible risks whilst actually allowing them to continue their work with little restriction. Such views continue to be expressed 30 years later but perhaps by somewhat fewer people than in 1975. The scientists on the other hand believed this to be a genuine and sincere attempt to allay any public concern by exhibiting a responsible attitude to the technology, based on a willingness to observe clear guidelines. Whatever one's view of the conference, what is clear is that the discussion facilitated the development, by the National Institutes of Heath in the USA, of appropriate policies that related the containment of particular GM organisms to the risks of harm that those organisms might cause. The evaluation of each new gene transfer experiment led to a numerical score and on the basis of that score the level of precautions and particularly of containment were determined. Such an approach then formed the basis for regulatory frameworks in several other countries, including the UK, in which such work was in progress.

There has been some relaxation of the guidelines over the past 30 years as it has been shown that in many situations the actual risks are much less than originally envisaged. However, as typified by the UK, many countries in which GM work is carried out still operate a clearly structured set of procedures for regulating this work. Local committees must be set up in any organization in which GM work takes place and these committees must have representation from the non-scientific personnel. Proposals for each new line of experimentation, including risk assessments and details of containment, must be presented to the local committee for approval. In making these decisions, the local committees are answerable to the Government's Health and Safety Executive (motto: *Reducing Risks, Protecting People*), which, if an organization fails to observe the national regulations,[2] can order particular lines of research to be stopped. Some of the regulations are much stricter than many people suppose. For example, those of us working with GM plants must ensure that their seeds do not escape to the external environment, however benign the plant and whether or not the genetic modification in question makes the plant more competitive.

At this point it needs to be said that one of the criticisms of the Asolimar conference was that ethicists were not invited and that there was little consideration of ethics. It is certainly true that there was no focus on possible intrinsic or deontological objections to GM technology but it is equally clear that there were concerns about risk. Risk has a clear ethical component, for example in considering whether it is acceptable to expose someone else to a risk that I am happy to live with, or whether it is justifiable to do something where there is some risk of environmental damage, however remote. The growth of GM crops provides a good example of the latter. In Chapter 6 we discuss in more detail the effects on biodiversity of growing particular crops. Here, we take the opportunity to look at more general risk-related ethical issues.

In relation to any release of GM organisms into the environment, typified by GM crops, opponents have often cited the precautionary principle. This comes in various guises but the version most widely used suggests that if an activity is believed to cause harm (and this instance we are talking about harm to the environment), then we should *not wait* for scientific proof of such harm before we cease/ban the activity in question. In relation to GM crops it is often coupled with the demand for proof that there is no risk. On the face

---

[2] At the time of writing this chapter, early in 2005, possible changes to the containment regulations have been put out to public consultation.

of it, these may seem acceptable positions to hold but further analysis reveals major difficulties. Consider the two following questions.

Questions

Can any activity be said to be risk free? Is it possible to prove absence of risk?

Starting with the first of these questions, we are hard pushed to think of anything that is risk free. All human activities, even lying in bed, carry some element of risk. Further, the answer to the second question is that we cannot prove a negative: we cannot prove that something does not exist. It is therefore logically impossible to prove the absence of risk. Thus it is very difficult to apply the this version of the precautionary principle or the demands about risks as set out by some opponents of GM technology. Indeed, if these ideas were widely applied, we would do never do anything new. We would simply stand still and that would be to ignore the risks of doing nothing.

So, if all human activities are at some level risky, how is risk to be evaluated ethically? This question may be answered at three different levels:

- risk and personal autonomy

- risk and other individuals

- general or widespread risks.

In discussing risks and personal autonomy we are thinking of the willingness or otherwise of a person to expose themselves to particular risks. In rich western society we encounter an anomaly here. Increasing affluence is linked with an increasing risk-averseness and an increasing willingness to go to law if we believe that we have been harmed by someone else's action or inaction, and yet many people participate in dangerous sports and other activities in which there is significant risk of injury or even death. In terms of personal activities it is a matter of an individual's own risk–benefit analysis, a form of personal consequentialism. I may choose to climb extremely severe routes or to canoe on grade 5 or 6 waters because of the positive feelings that I get from such challenges. At a more mundane level, I may regularly take journeys by car – not to do so would be inconvenient – despite the fact that about 10 people per day are killed in traffic accidents in the UK. I may choose to use my mobile phone frequently, although there is some evidence (albeit fragmentary) for radiation damage caused by long-term frequent use.[3] In this example, the tension inherent in risk–benefit analysis is even more apparent: I may like to use my mobile phone but I may not wish to have a phone mast in my village because of the risks of exposure to radiation. A final example, from the world of medicine, illustrates the whole topic very clearly: do I agree to have a treatment that is known to be risky but that, if successful, will effect a cure/enhance my quality of life?

How far then should an individual's willingness to accept particular risks for him- or herself allow other individuals to be exposed to risk? Some cases seem clear cut. If I choose to drive along my village street at 50 mph (80 kph), not only am I breaking the law but I am increasing significantly the likelihood that someone else may be injured or killed. Such an action that ignores consequences for others cannot be regarded as virtuous. But what about the yacht skipper who chooses to sail in storm-force winds and mountainous seas because of the thrill of meeting the challenge? His personal choice may cause the life-boat crew to be exposed to great risk, if he needs to be rescued. Or, particularly poignantly, should I consent to gene therapy treatment for my child, treatment that may cure his or her serious genetic condition but in which there are significant risks of long-term side-effects?[4]

In all these areas, it is clear that different people find different risks and levels of risk differently acceptable. However, in general, risks that are chosen are more acceptable than risks that are imposed. This leads us to consider risks that affect or are perceived to affect not individuals but groups or even populations. In general, these will be risks that are imposed externally, over which people have little or no control. Thus, returning to an earlier example, we can see that individuals may choose to use their mobile phones whilst being unwilling for the phone company to build a mast near the local school. The first is a matter of choice; the second is a matter of imposition. And it is this matter of imposition that brings us back to genetic modification: many were concerned and some protested about GM crops because of ideas about risks. It was claimed that unknown risks to human health and/or the environment were being generally imposed in the production of foods with little opportunity for avoidance. We know now that these fears are groundless: GM crops cannot of course said to be risk free, because nothing can, but they are no more risky than conventionally bred crops. Nevertheless, the debate was an excellent example of the tension between personal autonomy and public policy, a tension that occurs not only in relation to risk exposure but also in several other aspects of bioethics.

---

[3] For this reason, it is now recommended in the UK that mobile phones should not be used by children below the age of eight.
[4] See discussion of gene therapy in Chapter 7.

Possible misuse of genetic modification

It is a fact of human history that many of our inventions and discoveries have been used to harm others in crime and in warfare. Examples range from the use of natural poisons on the tips of arrows to destruction of complete cities with nuclear bombs. In view of this long and inglorious history there is no reason to suppose that genetic modification would not have the potential to be used in 'man's inhumanity to man'. Indeed, one of the topics that was raised in the Asolimar conference was the possible use of genetically modified organisms in biological warfare. Some commentators think that the natural range of organisms and biological materials is such that there is no need to use GM technologies to increase the range of available biological weapons. On the other hand, others have suggested that GM techniques may be used in the manufacture of more subtle biological weapons, for example, organisms that do not elicit any immune response or that are targeted at particular groups. In general, it was not thought necessary to make specific prohibitions about biological warfare. Biological warfare had been made illegal in the Geneva Protocol of 1925, which was updated in 1972, and a specific banning of the use of GM technology was not required. Nevertheless, there have been suspicions over the years that new GM biological weapons have been under secret development in particular government-funded laboratories in several countries, including the UK, the USA and, at least until the late 1980s, the former USSR. These allegations are of course without proof. As far as the regulatory frameworks governing GM techniques go, any experiment that involves genes encoding dangerous toxins or highly pathogenic organisms can only be done under the strictest of containment conditions in the specific context of research on human or animal health and with the specific permission of the appropriate committees.

5.3 Nano-technology

One of the most beautiful biochemical discoveries of recent years was made in the 1990s by John Walker and his research team at Cambridge. They demonstrated that particular components of the complex protein that makes ATP in the mitochondria[5] rotate like a turbine. Here then is a turbine-like rotor operating at the dimensions of a biological molecule. It is the development of technologies operating at molecular and even atomic scales that has been defined as nano-technology – technology at the nanometre scale.

---

[5] Mitochondria are the sub-cellular organelles that carry out energy conversions. The molecule known as ATP is the major energy-carrying molecule in a cell. John Walker and Paul Boyer (UCLA) were awarded the Nobel Prize for Chemistry in 1997 for their work on the mechanism and structure of ATP synthase.

Nano-technology is thus actually a whole range of technologies, some clearly biological or biotechnological but others not, that involve working at very small scales based around the nanometre, 1×10^{-9} of a metre. To put this into context, many proteins are several nanometres in diameter and the DNA double helix is just over 2 nm thick while the hydrogen atom is 0.1 nm across.

It needs to be said that the technologies being developed at this scale are all very much in their infancy but the many projected applications include the following.

- Further miniaturization – it has been suggested that writing with atoms would allow all 24 volumes of the *Encyclopaedia Britannica* to fit on a pinhead. Why one might do this is another matter, but more potentially useful examples of miniaturization include building a transistor from a single carbon atom and the development of a tiny protein motor based on the ATP-synthesizing enzyme that was referred to earlier.

- Development of drug delivery packages of sub-cellular size that can target particular cells within the body.

- Development of tiny robots (nanobots) that can travel in the blood stream to particular sites in order to carry out therapeutic processes that may include tissue repair, destruction of pre-cancerous cells before they become malignant or removal of harmful deposits from arteries. One suggested application is that NASA astronauts on a mission to Mars (possibly scheduled for the year 2020) may be implanted with nanobots that seek and destroy illness from within.

Our readers may think that all of these applications lie more in the realm of science fiction than reality. Indeed, Richard Fleischer's 1966 film, *Fantastic Voyage* (based on the book by Isaac Asimov), comes readily to mind. In this film a surgical team is miniaturized and travels in tiny submarine in a patient's bloodstream in order to locate and disperse a blood clot in the brain. However, we are not dealing here with science fiction: the applications listed above, along with many others, are the subject of extensive and active research. Nearly all the major developed nations are investing significant sums of money in nano-technology, expecting to reap benefits in fields as diverse as micro-manufacturing, information technology and medicine.

Interestingly, in the UK, the Royal Society organized a public consultation on nano-technology in 2003–04. The motivation for this was to avoid the situation that occurred with GM crops, where the public felt that they had been kept in the dark about the developments and suddenly found the products in their shopping baskets. Of those surveyed, 29 per cent said they had heard of nano-technology but only 19 per cent could make any attempt at describing what it is. Ethical concerns were general rather than specific: would the new technology present any as yet unknown risks; could it be misused in any way?

The public view coincides with much of the ethical analysis of nano-technology. At first sight, the applications mentioned above do not seem to raise any new ethical issues. Indeed, the two major ethical issues that come to mind are very old ones. First there is the issue of inequalities of opportunity. If we take medical applications as a specific example, it will be a question of who benefits from the new technology and based on long-standing precedent we know that it will be those who can afford it. The wealthy European, Japanese or American patient benefits from removal of artherosclerotic deposits in his or her arteries while in Africa tens of thousands of people die each day from hunger or disease. Second there is the issue of misuse and particularly misuse aimed at harming people. If a nano-technological device can deliver a therapeutic drug in a very precise manner it can also deliver a deadly toxin. As with genetic technologies, there is a need for a strong regulatory framework alongside existing international protocols.

However, it was not these age-old problems that raised the most publicized ethical concerns about nano-technology. Rather it was the possibility that the technology might 'escape' and even become uncontrollable. The first and possibility more realistic scenario is that nano-technology will result in the presence in the environment of millions of nano-particles with possible risks for human health, for example *via* effects on the respiratory system. The micro-particulate matter emitted from even the cleanest of diesel engines is cited as a precedent for the generation by human activity of harmful particles. However, supporters of nano-technology point out that many millions of tiny particles already circulate in the atmosphere, with no indication that they present a generic risk for human health.

The second scenario, presented by environmental campaigners, which in the UK included Prince Charles, is much more headline grabbing and has thus featured in scare stories in the press. It is that self-replicating nanobots will escape and then replicate in an unregulated manner, as with the self-replicating buckets of water in Walt Disney's film *Fantasia*. The result, according to the campaigners, would not be floods (as happened in *Fantasia*) but a mass of 'grey goo'. The likelihood of this happening is impossible to ascertain, although again it sounds like science fiction. Proponents of nano-technology suggest that self-replicating nanobots, if indeed they can be made, are very unlikely to be robust enough to replicate outside specifically controlled environments; they further suggest that those who raise such scenarios are simply anti-technology and will try any tactic to alarm the public. The campaigners, or at least some of them, are adamant however that there is a real danger and that, at a very minimum, there must be tight regulation of the way that nano-technology is developed and used. The next 20 years could prove very interesting or very alarming, depending on one's viewpoint.

5.4 Cybernetics[6]

For many people, perhaps including some of our readers, the term *cybernetics* is a topic more relevant for science fiction than for a book on bioethics. After all, cyborgs, highly realistic humanoids that are actually machines run by very sophisticated internal computers, have long been a favourite subject in science fiction. Amongst many films, *Blade Runner, AI* and *Terminator* are prime examples. However, in fact, cybernetics is a fast-growing branch of science in which, as several commentators have said, we are rapidly moving from fiction to fact. That is not to say that a cyborg is just round the corner. Nevertheless, very rapid progress is being made in interfacing biological systems with mechanical and electronic systems and it is this interfacing that gives us a working definition of cybernetics.

Mechanical–biological interfaces have been around for a long time: think of the traditional Long John Silver image of the sea-faring man with a wooden leg or of those early wooden false teeth. Of course, mechanical prostheses, replacing limbs lost in accidents or as a result of developmental defects or illness, have progressed a long way since the wooden leg. Further, we are able to use implanted devices that respond to signals from the body in order to regulate its activity, for example in heart pace-makers. However, in cybernetics the mechanical–biological interface uses the power of silicon chip technology to effect a more sophisticated two-way feedback between the prosthesis or implant and the nervous system of the organism. The possibility of such two-way communication was shown in the 1990s when functional connections between mammalian neurones and transistors were first made. Since then, progress has been very rapid. The demonstration that signals in the optic nerve of a cat could be transformed into a rough image of what the cat was seeing have led to several different types of human–silicon interfaces, including artificial retinas, that have given at least rudimentary vision to people with different forms of blindness. Even more amazingly, patients with locked-in syndrome[7] have been able to control, *by their thought processes*, a cursor on a computer screen *via* brain implants, which provided an interface between the patients' brains and the computer software. There are also, as was mentioned very briefly above, developments along these lines for wearers of prostheses. There has been a marked increase in the sophistication of some of these devices in recent years, raising the real possibility that they may be directly interfaced with the motor neurone network to achieve a much greater

---

[6] Much of the information used this section comes from material assembled by Christopher Hooks and Philippa Taylor, to whom we are grateful.

[7] In this syndrome, the patients' mental abilities and processes continue to function but they are unable to communicate or move because of damage to a specific part of the brain. It is a rare condition and, unfortunately for the patients, may be wrongly diagnosed as the permanent vegetative state.

degree of control. Further, the development of silicon chips that function more like nerve cells will increase the sophistication of implants that replace lost neuronal function.

However, it is not just in relation to restoring lost function that cybernetics is being used. Some computer scientists are developing interfaces that enhance human capabilities. Virtual reality programs in which the user immerses him or herself in another world, often by wearing a special helmet, are simple examples of these applications. One of the most enthusiastic proponents of cybernetic technology, Steve Mann at Toronto in Canada, has been developing wearable computers that look rather more like normal clothes than do virtual reality helmets. These wearable computers provide the wearer with a direct interface between his or her thought processes and computer networks. As Mann himself states[8]

> *Every morning I decide how I will see the world that day. Sometimes I give myself eyes in the back of my head. Other days I add a sixth sense such as the ability to feel objects at a distance . . . Things appear differently to me than to other people. In addition to having the Internet, massive databases and video at my beck and call most of the time, I am also connected to others. While I am . . . shopping, my wife, who may be at home or in her office, sees exactly what I see and helps me pick out vegetables. She can imprint images onto my retina while she is seeing what I see.'*

In the UK, the most enthusiastic 'cybernaut' is Kevin Warwick at the University of Reading.[9] He goes as far as saying that by developing human–computer interfaces we can achieve the evolution of the human species to a higher level. Rather than going down the route of wearable computers, Warwick has been developing micro-chip implants that enhance a person's capabilities. In a much-publicized experiment in 1998 he had a chip inserted into his arm that allowed him to be recognized so that he could switch on lights or open doors in his department. However, this was really little more than the internalization of information that might have been stored in a chip on his university identity card. The light switch operated or the door opened because a security screening device had recognized the information in the chip. More recently, however, he has achieved a simple form of communication (transfer of electronic signals) between himself and his wife *via* silicon chip implants that they each had. This is still a long way from being able to see the tomatoes in the supermarket as we described above for transfer of visual experiences *via* wearable computers, but for Warwick this is just the start. He envisages implants of tiny computers into the brain to provide the same range of capabilities as Steve Mann's wearable computers.

[8] Mann, S. (1999) Cyborg seeks community. *Technology Review* 102, 36–42.
[9] See www.rdg.ac.uk/KevinWarwick/html

Question

What ethical issues are raised by cybernetics?

The question is not easy. It typifies the situation described by Margaret Killingray

> *Sometimes our problems with knowing what is wrong and what is right arise because our world is changing so fast that we are constantly facing new situations that do not fit into our existing ways of thinking.*[10]

On the one hand, there is something very impressive about the possibilities that stroke victims and other paralysed people may gain some movement by thought transfer processes channelled *via* implants in their brain, or that wearers of prostheses may achieve much greater control or that the blind are made to see and the deaf to hear. All this is very much in accord with the beneficent use of technology to relieve human suffering. On the other hand there are aspects that, despite the enthusiasm of people such as Kevin Warwick and Steve Mann, seem rather more sinister, eliciting, if not a full-blown yuk response, a feeling of unease. How far should we go in enhancing an individual's capabilities to levels well above the norm? It might be argued that we have long used technology to enhance individual capabilities. We use telescopes and binoculars to see farther, we use infra-red sights to see in the dark, we use GPS for accurate navigation and we can send 'instant' pictures *via* our mobile phones (the 'everyman' version of viewing the vegetables in the supermarket). Are all these things any different from enhancing our capabilities through an implanted micro-computer?

Let us try to unpack this a little more. We suggest that there are three areas for concern. The first is that the technology will certainly only be for the few, those who can afford it. If the technology develops as enthusiasts wish it to, there will be a select group of people who have enhanced their ability to interact with the world and perhaps with each other in a vary dramatic way. It will increase the power of the wealthy over the less wealthy. Second, proponents of the technology such as Warwick envisage that eventually the thought-to-machine transfer process that has already been achieved *via* brain implants or the communication of sensory signals between individuals *via* wearable computers will eventually lead to implants that allow transfer of thoughts between individuals, thus setting up networks of enhanced humans that can communicate with each other in a special way. Maybe it sounds like science

---

[10] Killingray, M. (2001) *Choices*. BRF, Oxford.

fiction, and indeed 10 years ago we would certainly have held it to be so. However, based on recent progress, such developments must be regarded as real possibilities. The potential for mis-use is clear: a small network of people with enhanced capabilities able to communicate by thought transfer would certainly be in a position to exert power over others. And this leads to the third area of concern. If person-to-person thought transfer *via* implanted micro-computers does become a reality, might there not be the possibility of controlling the thoughts of other people, even of people who have participated willingly in the technology? The words of the rock group Pink Floyd, in their 1979 song *Another Brick in the Wall* come to mind:

> *We don't need no thought control.*

Certainly, the possibility of thought control is a very sinister one that alerts us to the need once again for clear and rigorous regulatory framework in cybernetics as in genetic modification and nano-technology. Perhaps in this instance the aspirations of the enthusiasts have given advance warning about what society needs to focus on.

6 Applications of genetic modification

Biotechnology presents us with a special moral dilemma, because any reservations we may have about progress need to be tempered with a recognition of its undisputed promise.

From *Our Posthuman Future*, Francis Fukuyama (2002)

6.1 Pharmaceuticals

The advent in the early 1970s of genetic modification (GM) techniques was quickly followed by their uptake into the pharmaceutical industry. Indeed, once the regulatory frameworks were in place (as discussed in the previous chapter) the pharmaceutical application of GM was very rapid and, even more than 30 years on, the pharmaceutical industry is still the major commercial beneficiary of the technologies. The first and best known example is the development of recombinant human insulin – most of our readers under the age of 35, in whatever country, will have been introduced to this example while still at school, so embedded is it in the 'story' of GM. One of the most intriguing factors in this was the very short time that elapsed between laboratory bench and clinic. The isolation and cloning of the gene encoding human insulin was reported in the scientific literature in 1977. At the same time, work was in progress that led to the expression of the gene in microorganisms and the patenting of this process. All three phases of clinical trials then followed and in 1982 the product was licensed for use in human therapy in the USA (and soon after in much of the rest of the world).

The advantages of human insulin for treatment of insulin-dependent diabetes are threefold. First, some people are allergic to the animal-derived (mainly pig) insulin that had been previously prescribed. Indeed, one of us has a diabetic colleague who is so allergic to animal insulins that he contends that the availability of human insulin has saved his life. Second, because production of human insulin is now controlled as an industrial process, its supply can be regulated in relation to demand. This is important in a world in which people are living longer (and therefore needing treatment for longer) and in which the number of cases of insulin-dependent diabetes is increasing (as a function of increased population). Third, quality control is easier in a regulated process involving large-scale growth of micro-organisms than it is in extractions from pancreases of slaughtered animals. So great are these advantages that human insulin is now the treatment of first choice in insulin-dependent diabetes, although it has proved unsuitable for a very small number of patients who had previously done well with animal insulins. However, the number experiencing problems with human insulin is a fraction of those who are allergic to animal insulins.

Dilemma

You are a young adult, newly diagnosed with insulin-dependent diabetes. After a range of tests your daily regimen of insulin injections is established and you are doing well on it. However, after some weeks you realise after reading the label on one of your batches of insulin that you have been prescribed recombinant human insulin. Because of your particular view of the natural world you have an intrinsic objection to all GM techniques, in whatever context. You ask your doctor to prescribe pig insulin instead. She refuses, stating first that it is not in your best interests to have pig insulin and second the very scarce supplies of pig insulin are kept only for those who actually need it because they experience problems with human insulin. How should you proceed?

This success with human insulin was followed rapidly by the development of other recombinant pharmaceutical products, including human growth hormone, anti-viral drugs, drugs for cancer treatment, and many vaccines. Indeed, it is likely that many of our readers will have received a 'recombinant' vaccine of some sort. In all these cases, the product made by GM technology is regarded as safer, with less possibility of side-effects than the previously available version (if indeed, a version had been available at all: several of these products would have been so difficult to obtain, existing within natural sources in such short supply, that pharmaceutical application was out of the question).

Some of these aspects are well illustrated by human growth hormone (HGH). This has been used for some time in the treatment of children whose growth is restricted because they do not produce enough of the hormone. The previous source of the hormone was pituitary glands of dead people. However, after several years of this use, it became apparent that many of these hormone preparations were contaminated with the agent that causes the very distressing and eventually fatal neuro-degenerative condition, Creutzfeld-Jacob disease (CJD).[1] The use of pituitary-derived HGH was immediately banned. Claims for compensation against the relevant medical authorities (in the UK, the National Health Service) still arise as new cases of CJD are diagnosed even 15 years or more after cessation of treatment (the disease has, in some cases, a very long incubation period). As always in such situations, establishing a legal claim is difficult because the treatment was given in good faith in the patient's interest and none of the parties involved in supplying the drug (e.g., the pharmaceutical manufacturer, the clinician) could have known about the problem (indeed, why pituitary extracts are so frequently contaminated with the CJD prion, in relation to the extreme rarity of the disease in the population, remains a mystery). Nevertheless, at least in European countries with nationally provided healthcare, compensation has been awarded, usually on a 'no-fault' basis.

The arrival on the market of HGH made by GM techniques, and thus free from any danger of CJD, was therefore very welcome, and treatment of children with hormone-based growth restriction was resumed. However, the ready availability of supplies of uncontaminated HGH led to other uses. Firstly, there has been pressure from parents of children who although not growth restricted by under-production of HGH are nevertheless much shorter in stature than the population mean. In some instances, and especially in countries where medicine is on a more commercial basis, such as the USA, clinicians have responded to this pressure by prescribing HGH. This raises interesting questions about how individuals and society in general regard those who not fit with expected norms, questions that we examine more fully in the next chapter. Returning specifically to HGH, its ready availability also led to its use in sport. Supplies of HGH are available to those who are determined to find them and the hormone is taken by body-builders who wish to increase body mass. It has also been taken by participants in competitive sports including American football, track and field athletics and weightlifting. This is a classic case of the misuse of drugs in sport and is banned by the national and international bodies that administer sport. Nevertheless, as we have seen recently in the Athens Olympic Games of 2004, there are those

---

[1] The causative agent is a 'rogue' protein called a prion; CJD is thus in the same 'family' of diseases as 'mad cow disease' (bovine spongiform encephalopathy, BSE).

who will do anything to gain a competitive edge, without any regard for the regulations or for their long-term health (this must be a particular concern when HGH is administered by unscrupulous coaches to young people who have not yet completed their growth phase). Overall then, HGH provides several opportunities to think about how society uses the findings of science.

Question

Should misuse or misapplication of a technology lead us to consider banning it?

We suspect that many of our readers will think that this is a non-question. They will argue, with some justification, that, for example, recombinant HGH has brought great benefits and that to ban it would deny its use to those who need it. What is needed here is tight regulation to ensure that it is not misused. They may go on to give many examples of inventions that have brought great benefits but that, sadly, have also been used to do harm. Nevertheless, the harm versus benefit equation that we have encountered in several places already can make a valid contribution to ethical debate. The question then is whether the possible misuse of a technology so outweighs the possible benefits that attempts at a ban are justified.

The development of the pharmaceutical products of GM technology has gone on mainly away from public gaze. The majority of recipients of 'recombinant' products are probably unaware of the method of manufacture and indeed are probably not interested. For most, the main interest in pharmaceuticals will be based around questions such as 'Is a suitable drug/vaccine/ treatment available for me?' and 'Will it be effective?'. Indeed, such is the lack of awareness of the way that GM is embedded in the pharmaceutical industry that even some of those with a deep concern about other applications of GM technology have been unaware of the breadth of the range of GM-derived pharmaceuticals. The same is not true however for the application of GM in agriculture and the food industry, topics to which we now turn.

6.2 Food and crops

A brief introduction

For those of us who are observers of the science, technology and bioethics scenes, it is difficult to remember a topic that, in the UK and in some other

European Union countries, has generated as much opposition as GM crops. We examine the elements of the opposition to GM crops below but first it is necessary set out some definitions. The term *GM crop* is self-defining – *a crop variety that carries a genetic character as a result of genetic modification*. However, a GM crop is not necessarily a GM food, because products of GM crops such as oils, proteins or starch may well not have been modified by the GM process. Herbicide-tolerant oil-seed rape (canola) and soya bean fall into this category. A *GM food*, strictly speaking, *is a food where the genetic modification of the crop from which the food is derived was aimed at changing the composition of the food substance itself*; for example, the plant breeders may have used GM in order to change the biochemical composition of the proteins. No such foods are currently under consideration for commercial growth but may come on stream in the future. However, that is not all. Confusingly, there are *intermediate categories* in which the new genetic material will be consumed with all the rest of the DNA or in which a new protein, modified or added for non-food reasons, is consumed with all the other proteins, when the plant material is eaten raw or cooked. The non-softening tomato, the first GM crop to reach the market (see below), is an example of the former, while insect-resistant sweet corn would be an example of the latter (unless the plant breeders had ensured that the insect-resistance gene was switched off in seeds, although of course the new DNA – but not the new protein – would still be consumed).

Some thoughts on cheese

In the initial phases of manufacture of several types of cheese, milk proteins are partially digested and the milk clotting is initiated by the action of a mixture called rennet, which is obtained from the lining of calves' stomachs. The main active component of rennet is an enzyme called chymosin, and during the 1980s the cow gene that codes for this was isolated and cloned; the gene was then transferred to micro-organisms so that they produced active chymosin. This is of course directly parallel to the production of insulin, discussed above, although a different type of micro-organism was used. It was a short step from there to show that recombinant chymosin was effective in cheese-making, and in the UK cheese made this way went on sale in the early 1990s under names such as 'vegetarian cheddar cheese'. The description 'vegetarian' was used because no extracts of calves' stomachs had been employed in the cheese-making process.[2] Therefore, what we have here is a process that employs a product of GM technology directly in food production. Furthermore, this recombinant enzyme, or at least a significant proportion of it,

[2] Of course, this overlooks the fact that in countries where there is little demand for veal many of the male calves are killed anyway.

remains in the final food product, although it is denatured (loses its activity) in the process. It is therefore consumed along with the other cheese proteins. However, the cheese proteins themselves were not modified by the GM process, so this is not a GM food in the strict sense but falls into one of the intermediate categories referred to above. It remains a fact that the recombinant protein is consumed with the cheese, but vegetarian cheeses made with recombinant chymosin remain on the market and have not attracted the type of opposition that we see with GM crops (see below). Of course, the concerns about environmental damage do not arise with this cheese-making process, but nevertheless it is interesting that, in the climate of the late 1990s and the early 21st century, these cheeses should have been accepted so readily.

Crop plants

Genetic modification of plants was first achieved, albeit with a very low success rate, in 1983 (although GM techniques were already being applied to the study of plant genes, as is described briefly in Section 6.4). The initial development of plant genetic modification relied on natural mechanisms, mainly based around the way in which a bacterium, *Agrobacterium tumefaciens*, transfers some of its genes to plants. Specific details of this methodology and the subsequent development of other techniques such as the 'gene gun' lie outside the scope of this chapter.[3] Here we deal with the main points in the debate:

- the advantages and disadvantages of GM techniques

- the applications of GM techniques in crop plants

- the growth of GM crops across the world

- the objections that are raised against GM and the answers given by proponents of the technology in the following areas:

  ○ possible intrinsic objections

  ○ risk

  ○ consumer choice

- a different approach to evaluating new crop varieties.

So, what are the advantages and disadvantages? The technique is both *precise* and *imprecise*. It is precise because, unlike 'traditional' plant breeding tech-

---

[3] Readers who are interested in this should consult the 'Suggestions for further reading' at the end of the book.

niques, one or a few specific genes, conferring desired characters, are transferred to a plant, the rest of whose genetic characteristics are not altered. Thus, single wanted genes can be moved with precision into particular crop varieties. This clearly contrasts with what happens in conventional breeding.

However, genetic modification is imprecise because of *position effects*: there is no control over where in the plant chromosomes the incoming genes are inserted. This causes great variation from plant to plant in the first 'GM generation' in the level of expression of the new gene or genes. Therefore there is a need for selection of the first generation of GM plants followed by observation of the stability of inheritance in subsequent generations. However, in practice, the overall process is still faster than the sorting, re-crossing and re-selection that has to occur in conventional breeding.

Thus the advantages of GM are that specific genes may be inserted into well characterized varieties (without bringing in 'unwanted genes'), that these genes can come from widely separated species and that new varieties may be more quickly adopted for use.

As with the original techniques of GM in bacterial cells, initial development of crop GM was very rapid, and by 1985 strictly regulated field trials of GM crop species were taking place in several countries.

Question

Why might there be more general concern about conducting trials with GM crop plants than with GM micro-organisms?

The most obvious answer to this question is that, unlike micro-organisms, crops grow outside and are therefore not contained. If GM does in some way make crops more invasive or if the genes that have been inserted confer unexpected advantages to wild species should hybridizations occur, then the uncontained growth of GM crops would be a very risky business. For that reason, the development on a commercial scale had to be preceded by careful evaluations of hybridization rates, gene flow and ecological competitiveness. Nevertheless, commercial-scale field trials, mainly but not exclusively in the USA, were under way by the early 1990s and the first products of GM crop technology went on the market in the mid-1990s. Currently, the major GM crop is herbicide-tolerant soybean, which was rapidly adopted in US agriculture. Between 1996 and 2002, the area devoted to this crop in the United States increased from 1.7 million to 34 million hectares. This represents about 70 per cent of the US soybean crop. On a worldwide basis, in 2004, some 81 million hectares spread amongst 17 countries were used for GM crops, including Argentina, Australia, Brazil, Canada, China, India and South Africa and of course the USA. Further, China has just approved the commercial cultivation of a GM rice variety. Of the five million farmers in the world currently growing GM crops, 75 per cent are small-holders in China, and many of the users in South Africa also farm on a small scale. Indeed, according to the International Service for the Acquisition of Agri-Biotech Applications,[4] 90 per cent of the farmers who grew GM crops in 2004 were resource-poor farmers from developing countries, whose increased incomes from biotech crops contributed to the alleviation of poverty.

The main traits – genetic characters – transferred by GM techniques are herbicide tolerance, resistance to insect pests and resistance to viruses. Traits involving crops not destined for human consumption include the production of edible veterinary vaccines in maize (corn). Many other applications are being developed.

Questions

What are the key points against and in favour of the use in agriculture of GM plants?

What types of ethical argument are you using in presenting these points?

Although, as is evident from the data presented above, GM crops have penetrated agriculture in several parts of the world, they have not, at the time of writing (late 2004) done so in countries of the European Union, although there is some small-scale commercial growth of GM maize varieties in Spain while approval has been given for growth of certain GM crops in Portugal. This slowness in adopting GM as an approved method of plant breeding has been caused partly by caution in the European Parliament and in the parliaments of individual EU countries and partly in response to vigorous campaigning by organizations opposed to the growth of GM crops. In the UK, and in other countries of the EU, the intensity of this campaign increased very markedly in 1999. It was accompanied in the UK with sensationalist press reporting, using terms such as *Frankenstein foods* and *frankenfoods* that hinted at mad scientists losing control of the products of their ill conceived work. The campaign did not lead to a ban on the import of derivatives of GM crops (such as soya products), but it did cause the withdrawal from the market of tomato purée made from GM tomatoes in which one of

---

[4] www.isaaa.org

the tomato's own genes had been turned back to front in order to delay softening. It also upped the stakes in respect of the both industrially and government-funded farm-scale trials of GM crops, and in several of the former, protestors ripped the GM plants from the ground. This was and still is a subject that raises strong feelings. Furthermore, the main organization that validates organic agriculture in the UK, the Soil Association, has declared GM crops 'non-organic'. There is a different situation in the USA, where different organic farming groups take different positions on GM crops.

The key points made by those *opposed* actively to GM crops fall into four groups:

- intrinsic objections
- risk
- consumer choice
- wider social issues.

The fourth group of issues is dealt with in the next chapter. Here we deal with the first three.

Intrinsic objections

As we saw in Chapter 5, there are some opponents of GM technology who believe that moving genes between organisms is intrinsically wrong. This view was first expressed when genetic modification of bacteria was developed in the 1970s, but little more was heard of it until the widespread use of GM crops became a real possibility. In the UK some of the most ardent and vocal of the anti-GM campaigners hold this view, which some regard as being almost religious in character, and for them it is important to resist as far as possible the use in the UK of these crops. Even though this is likely to be a very minority view in wider society, we still need to consider the problem of how to make provision for people who hold this position.

Risk

Risk and the precautionary principle were discussed in the previous chapter. In this chapter we show how these relate to the debate on GM crops. The main risk factors raised by those who oppose GM crops are the following.

- *Containment.* As we have already noted, crop plants, by their very nature are not contained. There are already millions of hectares of land devoted to GM crops. Some have described this as letting the genie out

of the bottle, on the grounds that GM crops present threats that conventionally bred crops do not.

- *Marker genes.* Maker genes are those added, along with the gene of
  interest, in order to identify the cells that have been successfully genetically modified. The first commercial GM crop varieties contain antibiotic-resistance marker genes and concerns have been expressed that
  there is a chance, albeit very remote, that such genes may find their way
  from the crop plant to bacteria that infect humans or farm animals.

- *Gene flow and superweeds.* Will the GM crop be able to cross with a
  related species that grows in the wild, thus allowing movement into the
  wild species of genes that may change its character, with the possibility that it may be transformed, for example, into an aggressive weed?
  Another possibility suggested is that the GM crop itself may be able to
  establish itself as a 'superweed'.

- *Biodiversity and sustainability.* It has been suggested that several of the
  traits introduced into crops by GM techniques threaten biodiversity.
  For example, the herbicide treatment regime applied to herbicide-
  tolerant crops may reduce weed populations to such an extent that there
  is a negative effect on organisms ranging from insects to birds that
  depend directly or indirectly on the weed species for food. Another
  example raised by opponents of GM crops is the possibility that genes
  conferring resistance to insect pests may, by virtue of their presence in
  pollen, have an adverse effect on beneficial insects. Use of GM crops
  therefore, it is said, is incompatible with farming sustainably.

- *Food safety.* It has been suggested that the presence of foreign genes
  may cause the synthesis of unknown by-products or of new allergens.

Consideration of all these factors leads to the strongly held view that the precautionary principle in its 'strongest' form (see Chapters 3 and 5) should be
applied. This is essentially a consequentialist argument: there are risks that
the consequences of growing GM crops may turn out to be disastrous. These
risks, however theoretical they may seem to be, mean that we should not go
ahead with the adoption of GM crops into agriculture (meaning specifically
European Union agriculture, since it is a little late to prevent their use in, for
example, the USA). Indeed, a recent 'battle-cry' from campaigners in the UK
is that they must seize the chance to prevent commercial growth of GM crops
while there is still time.

It needs to be said that many scientists were very surprised at the strength
of the opposition to GM crops. Some suggested, perhaps arrogantly, that
opposition was based on ignorance, but others recognized a strong ethical

element in the arguments, even if they did not agree with them. Of course, those who *support* the adoption of GM crops in EU agriculture have their corresponding counter-arguments (although they will recognize that there is no counter to intrinsic or deontological objections) as follows.

- *Containment.* Supporters of GM technology maintain that there is no evidence that GM crops are any more dangerous than other crops and there is therefore no need for special containment. They cite data from long-term experiments on gene flow and on the behaviour of GM and other crops in the field, from the UK's farm-scale evaluations and from the large-scale commercial growth in other countries.

- *Marker genes.* It is accepted that there is a very remote chance of antibiotic resistance genes being transferred to bacteria, but it is contended that the risk is so small that it cannot be quantified. Nevertheless, those involved in the development of GM crops acknowledge the importance of ensuring that, where these antibiotic-resistance marker genes are used, the antibiotic in question should not be one that has applications in human or veterinary medicine. Furthermore, methods, albeit not straightforward, have been developed for removal of marker genes, and in any case more recently developed GM crops contain different types of marker.

- *Gene flow and superweeds.* Supporters of GM crops point out that all crop varieties are selected for growth and yield in agricultural systems and in general they perform poorly in the natural environment. Although it is true that some species can grow as 'volunteers' in the wild, they do not become established as ongoing populations. Data to support this view have already been alluded to above. Nevertheless, it remains possible that a new genetic trait may make a crop species a better competitor so that it threatens native species or becomes a nuisance. However, the risk, if there is one, is related to the trait itself and not to the method used for introducing it into the crop.

  Gene flow, the movement of genes from one crop variety to another or even from the crop to a related wild species, is another point raised by opponents of the technology. In reply, supporters of the technology point out that in the UK there are very few crops that are capable of out-crossing readily with wild species. However, two of them, beet and oil-seed rape, are amongst the crops for which GM techniques have been developed. There have been some very extensive studies of the formation of hybrids between oil-seed rape and wild mustard and wild radish; these hybrids occur at very low frequencies but there is no evidence that they become established in the wild (readers are however reminded of the discussion about risks and proving negatives in Chapter

5). However, scientists point out that GM crops are no more or no less likely to outcross than conventionally bred crops, and thus the consequences of any gene flow depend on the genetic trait itself and not on the breeding method.

- *Biodiversity and sustainability.* A moment of thought will tell us that the aim of agriculture is to produce food and application of that aim will lead a farmer to limit as far as possible the growth or activity of other living organisms that may compete with or damage the farmed crops (and animals). A farm, whether conventional or 'organic', is not a natural ecosystem. It is acknowledged that in developed countries much of agriculture is practised as an intensive, almost industrial system. Indeed intensive agriculture, combined with very focussed breeding programmes, has contributed significantly to food security in developed countries. However, even the most ardent proponents of intensive agriculture recognize that it has had very marked effects on the landscape, the land itself and wildlife. Indeed, one very clear outcome of the UK's farm-scale evaluations of GM crops (the results of which were published in 2003) is that we are much more aware of effects of agriculture itself. This is a classic utilitarian trade-off: loss of biodiversity versus production of food for people. A 'back-to-nature' approach is not workable. Nevertheless, within the farming and agricultural science communities there is now an acceptance of farming more sustainably – lessening, as far as is compatible with production of affordable food, the effects of farming on the environment. Supporters of GM technology in agriculture suggest that GM may actually help in the move to greater sustainability by the production of new crop varieties better suited to a more sustainable approach.

- *Food safety.* Are the fears about food safety raised by some of the opponents of GM crops actually well founded? Certainly, based on surveys and opinion polls, and more locally, in the experience of one of the authors of this book, it is apparent that members of the public were and possibly, early in 2005, still are concerned about food safety. However, regulatory authorities in several countries have found no evidence at all that the technique of GM itself raises any food safety issues, although both these authorities and supporters of GM crops can envisage situations where particular genetic traits, introduced by GM, may cause problems (so, for example, genes encoding potentially allergenic nut proteins should not be transferred to plants in which their presence would not be expected). Supporters of the technology also concede that plants containing a foreign gene that modifies a metabolic pathway must be subject to rigorous analysis as if they were completely novel crops (because of the possibility of unusual by-products). Overall however, as proponents of GM are keen to point out, after nine years of large-scale growth in the USA (and more recently in other countries) there has not been a single instance of a food safety problem with the GM crops in current production, a fact even acknowledged now by many of the opponents of the technology.

Risks of GM crops – a different approach

The arguments and counter-arguments set out above give a picture of two sides, entrenched in particular positions, lobbing verbal and written grenades at each other, and to some extent that picture is true within the European Union, and especially in the UK.[5] Indeed, the continued attempts of supporters on both sides of the debate to score points off their opponents has actually made it very difficult for the public to discern what the real issues are. However, over the past two years or so, a new approach is emerging that it is hoped will help to achieve a satisfactory outcome to the debate that recognizes on the one hand the good science that lies behind GM technology and the potential of the technology, and on the other hand the concerns of the campaigners and the wider public.

The basis of this approach is that GM technology should be regarded simply as an addition, albeit a very useful addition, to the plant breeder's tool kit. Indeed, it is argued that GM techniques may be less invasive than other methods used routinely in plant breeding, such as forced hybridizations or the induction of mutations by exposing seeds to ionizing radiation. The latter methods were used in the breeding in the UK of the popular malting barley variety, *Golden Promise*, and in the wider world of several of the higher yielding cereals produced during the Green Revolution. It has been a mistake, it is argued, to focus on the technique for introducing genes into plants rather than on the use to which the technique is put. This focus has led to crops bred by GM techniques being regarded as almost a different class of crops; it is as if we had designated those varieties bred by the induction of mutations as '*mutant crops*' – one wonders what the public reaction might have been. Overall, a change of focus from the technique to the genetic trait[6] has found favour with the scientific community and with some of the opponents of GM crops. However, there remain some campaigners either with intrinsic objections to all genetic modification or who regard GM techniques as being so far from what they regard as natural that they are unconvinced by this approach.

---

[5] There has been opposition to GM crops in other countries, including the USA (some of which was initiated by visiting campaigners from Europe), but it has been less organized and very much less successful.

[6] A fuller outworking of this approach is presented in the relevant publications listed under 'Suggestions for further reading' at the end of this book.

Therefore, if we regard GM as a plant breeding technique, what should be our concerns; what should be evaluated? As has been hinted at in the discussion above, what really matters when it comes to environmental or food safety is the genetic traits of the crop, not the method by which the traits were bred into the crop. This approach was nicely illustrated in the UK by the farm-scale evaluations of three herbicide-tolerant crops bred by GM techniques. In the trials the GM-bred strains were not compared with herbicide-tolerant strains generated by non-GM techniques (there are none such yet in commercial use but they are under development). The comparison therefore was with non-herbicide-tolerant but otherwise similar strains. It is immediately obvious that what was being tested in these trials was the herbicide-tolerance trait and the associated system of crop husbandry. In all three more efficient weed control was achieved, but with two of them there was some above-ground reduction in biodiversity[7] in and immediately around the fields of crops (no attempt was made to study below-ground biodiversity), while with the third, a herbicide-tolerant maize (corn), there was a slight increase in biodiversity. Genetic scientists may well join with the anti-GM campaigners in suggesting that the trials have told us very little. However, they do indicate that GM as a technique should not be a specific target of concern. It is what is done with it (or indeed with any other plant breeding technique) that should be evaluated.

Consumer choice

The products from GM crops have already achieved widespread penetration of world food markets (and/or other markets, such as that for cotton). A very significant proportion of the world's soya oil and protein comes from genetically modified herbicide-tolerant plants (but of course this genetic trait does not affect in any way the composition and quality of the oils and protein). Further, it is likely that the European Union and several of its member countries, including the UK, will soon give approval to grow certain GM crops while the area under such crops in other countries continues to increase. In order to help identify products of GM crops, labelling protocols have been adopted in the hope that consumers will be able to choose or reject these products. Ethically this is a virtuous action, giving space to others to exercise their scruples. However, it turns out to be more complicated than at first thought. First, different countries have different labelling regulations. In the USA, for example, derivatives of GM crops that are themselves not affected by the modification do not have to be labelled. Examples of this are the oils

---

[7] Very recently published results from a different series of trials indicates that if these crops are used in a crop rotation system so they are not grown in the same fields every year there is no reduction in biodiversity.

and proteins from herbicide-tolerant soybean plants: in the USA this distinction between GM crops and GM food is clearly made. In other countries, the unaltered products of genetically modified plants do have to be labelled with a clear indication of the source. The consumer is thus unclear as to what lack of a label may indicate.

Another difficulty is deciding how far along the economic chain the labelling should go. For example, will restaurants be required to state in their menus that certain items are made from the products of GM crops? At present, many make the claim that their food is GM free, whilst others with a different purchasing policy state that they cannot guarantee that their products are GM free. In this area of conflict between personal morality (and hence personal choice) and public policy difficult situations always arise and it is hard to predict how this one will be resolved.

However, there is one area of personal choice where appropriate courses of action can be taken. In countries such as the UK where GM-bred crops are regarded as non-organic (however they are grown), it is a concern to the organic farmer that the crop may be contaminated with non-organic material. So, for example, a farmer growing organic sweet corn[8] near to and down wind of a GM-bred maize (corn) crop will certainly note some pollination of his or her crop by pollen from the GM maize. The concern then would be that the percentage of cobs 'contaminated' in this way exceeds the limit permitted for validation as organic. Thus in the UK, where this attitude to GM-bred crops within the organic farming community looks set to continue, planting regulations for crop varieties bred by GM techniques must ensure that the livelihood of organic farmers is not threatened.

Summary of Section 6.2.3

- Plant GM was first developed in 1983, using a modification of a natural system that transfers genes to plant cells.

- After a period of extensive development, including controlled field trials, GM-bred crops came to market in the mid-1990s, and the area devoted to their growth in several countries, but especially the USA, has increased dramatically.

---

[8] Maize and sweet corn are the same species; sweet corn is a mutant that fails to turn much of its sugar into starch.

- In the UK and in most other EU countries, commercial growth of GM-bred crops has not yet been approved, but this may happen in the near future.

- Since the late 1990s there has been a concerted campaign in the EU and especially the UK against the growth of GM-bred crops. Some of the objections are based on a view that any form of genetic modification is intrinsically wrong but most of the arguments are consequentialist.

- Objectors believe that there are unacceptable risks to the environment and to food safety.

- Proponents of the technology hold that these fears of risks are unfounded.

- There is little common ground between the two sides in the debate.

- The UK and other EU governments seem to be convinced that GM-bred crops do not present any new risks.

- It has been suggested that a more helpful approach is to regard GM as a set of techniques in plant breeding and that when it comes to new crops (whatever breeding method has been used to create them) the new genetic trait should be evaluated rather than focussing on the breeding method.

- There are also issues of personal choice, which will be difficult to resolve.

Case study

This case study is aimed especially at readers in the UK but it does raise issues of a more general nature. It is based on real situations in a particular region of the UK. In the study, we assume that the European Union has approved GM crops in principle and that the UK government has given the go-ahead to grow certain GM crop varieties.

You are a district councillor in a region noted for tourism and for agriculture. One of the district's advantages is a mild winter climate, enabling farmers to produce some of the earliest crops in the UK as well as some more exotic crops. In common with other districts in the county, there are several farmers' markets and about 15% of local produce is claimed to be 'organic'. The county's commercial development agency has been working on 'branding' the whole county (including your district) and its products in order to increase market penetration. 'Local foods from lovely Loamshire' is their favourite slogan; the Lovely Loamshire designation and its logo are registered trademarks.

The council is about to debate a motion, proposed by a Green Party councillor and seconded by an independent councillor, that the district should ban the growth of GM crops. Your task, as the senior member of your party's group on the council, is to set out the issues in a clear way, as unbiased as possible, so that your group can discuss the issue at its meeting to be held before the council's debate.

Notes: (1) In addition to the issues related to the crops themselves, there are also issues about the personal freedoms of farmers who wish to grow the GM varieties and issues about the legality of banning a commercial activity approved by both national and European parliaments.
(2) A similar case study, with a fuller discussion of the issues, has been published by the Higher Education Academy's Bioscience Centre and may be consulted at http://www.bioscience.heacademy.ac.uk/resources/ethicsbrief.htm

6.3 Genetic modification of animals

Introduction

Genetic modification of animals actually preceded that of plant cells, but we have preferred to discuss the topics in order of increasing complexity of the organisms involved. Success with animal cells followed within a few years of the first experiments with micro-organisms and for some types of animal cell was well established by the late 1970s. Further progress was rapid and techniques for genetic modification of mammals (not just mammalian cells) had been developed by the early 1980s.

We deal with the topic here under three headings:

- the scientific background

- applications of animal genetic modification

- animal welfare issues.

Scientific background

There are two basic procedures by which this genetic modification in non-human mammals can be achieved. One procedure uses either the unfertilized egg (oocyte) prior to *in vitro* fertilization, or the newly fertilized egg immediately after entry of the sperm, as the target for the foreign DNA. If the embryo develops normally (and it needs to said that the success rate at this stage is much lower than in 'normal' *in vitro* fertilization), it is introduced into the uterus of a suitable potential mother in order to establish a pregnancy. The other procedure uses embryonic stem cells (see Chapter 9) as the target for the added DNA. The GM stem cells are then put back into the embryo. The resulting mammal is of course a mosaic – a mixture of engineered and non-engineered cells – but, for reasons that are not understood, the germ-line cells that develop within these (partly) GM animals often carry the 'foreign' gene and thus are a source of that gene for establishment of completely GM animals in the next generation. This method is especially useful for animals with short generation times such as mouse, but with larger animals such as sheep and cattle adding the foreign DNA to the oocyte is the method of choice.

The genetic modification of mammals is thus now an established practice and furthermore techniques have been developed that ensure that the foreign gene is active in the right cells/tissues at the right time. In other words, by making sure that the foreign gene is joined to an appropriate promoter ('on–off switch') it is possible to control the gene's pattern of expression.

However, despite the overall success of mammalian GM techniques, there are still some problems. First, success rates from *in vitro* fertilization are much lower with GM embryos than with non-modified embryos. Second, and similar to the situation in plants (described in Section 6.2), the level of expression of the foreign gene varies considerably between different individuals. This is again mainly because of differences in the place in the chromosome at which the foreign gene inserts (although it is likely, based on recent research, that targeting genes to locations within the chromosome where they may be highly expressed may soon be possible). At present therefore it is necessary to select the GM animals with the highest rate of expression, and since some of the target species are large farm animals such as sheep and cattle there may be only a few from which to select. Third, even when a 'high-expressing' animal has been selected there is no guarantee that the foreign gene will be equally highly expressed in later generations.

Question

Does genetic modification of mammals raise any ethical issues that are not raised by genetic modification of micro-organisms or of plants?

Applications

Against the background of some technical difficulties, as mentioned above, it has proved possible to create GM mammals for use in medical and biomedical research and in biotechnology. Indeed, in the UK, many hundreds of thousands of GM mice are used every year in research on specific diseases. For example, they can be modified so that they become experimental models for studying human genetic diseases, including cystic fibrosis and Huntington's disease, or they may be modified with oncogenes (genes that when activated cause the animal to develop cancer). Also well established is the creation of GM farm animals such as sheep that produce pharmaceutical proteins in their milk. Another current research line is the modification of pigs so that their organs may be used in human transplants.

Animal GM and animal welfare

We presented an extensive discussion of animal welfare in Chapter 4, but here we simply focus on those animal welfare issues that may arise from genetic modification. In the light of the earlier discussion we cannot avoid the conclusion that genetically modifying animals for biotechnology or for research on animal or human genes is an instrumental use of animals. Furthermore, in those experiments where the GM animals are models for human genetic disease, many of the animals suffer physical and physiological malfunction and in some cases actual pain as a result of carrying the mutant gene. Therefore, are we are justified in creating, for example, mice that develop cystic fibrosis or cancer, sheep that produce drugs in their milk or even pigs with 'human-friendly' organs?

Based on our discussion in Chapter 4, we think it likely that many people find it more acceptable to use animals in medical research than in factory farming or in testing of cosmetics. Many will argue along consequentialist lines, from a human-centred position, that, provided no *un-necessary* suffering is imposed, the use of animals in this way is justified by the benefits for human health and welfare. What, it will be said, is wrong with sacrificing tens of thousands of mice if it saves the life of one child with cystic fibrosis? Indeed, such a view will find support amongst many professional ethicists, philosophers and theologians. However, again as we saw in Chapter 4, there are significant numbers of people who believe that the use of animals in medical research can never be justified by the results. They reject the consequentialist argument, believing that it is intrinsically wrong to impose suffering on non-human animals (although some would impose this restriction only on research involving warm-blooded vertebrates, i.e., birds and mammals). As with the opposing view, those holding this view can also cite the views of particular ethicists, philosophers and theologians.

In the developed countries of the world the first of the two views, namely that using animals in medical research is justified, is the majority view (even though a very vocal minority can make their opinion known, sometimes by violent means). For those who take this majority view, genetic modification of animals does not seem to present any new problems that have not already cropped up in a general discussion of the use of animals in research. If the argument proceeds along consequentialist lines, there seems little or no difference between genetically modifying pigs so that their organs can be used for human transplants and breeding pigs for pork and bacon or between doing research with mice carrying human oncogenes and using mice for research on carcinogenic chemicals. Nevertheless, some thinkers have suggested that genetic modification may offend against the natural purposefulness of animals (the Greek word for this is *telos* and readers will recall the work of Aristotle in relation to ethics based on natural law, discussed in Chapter 2). For example, such thinkers may ask whether genetic modification of a sheep denies its 'sheepness'. Certainly a sheep that produces a human protein in its milk is making something that sheep do not naturally make. However, the sheep does not take on human attributes. It remains clearly a sheep, just as a bacterium making human insulin (Section 6.1) remains a bacterium. Most ethicists therefore hold the view that genetic modification of this type does not offend against the essential *telos* of the animal. If this is the case, then *GM itself* presents no new ethical problems in relation to humankind's use of animals, although there may be some *applications* that raise specific concerns.

Exercise

There is a general shortage of organs for use in transplants for human patients. Pig organs have been considered as suitable for transplantation into humans except for the major problem of immunological rejection. However, it is suggested that pigs may be genetically modified so that, immunologically, their organs resemble human organs, thus decreasing significantly the likelihood of rejection. Evaluate this suggestion ethically from the following standpoints:

Human healthcare
Animal welfare
Animal 'rights'
Your own religious or philosophical world view
Whether there are viable alternatives

Note: this is based on a more extensive exercise developed for the Salters–Nuffield Advanced Biology curriculum.

6.4 Research uses of genetic modification

Anyone looking at a molecular biology or genetics textbook in 2005 and comparing it with an equivalent text in 1975 might be forgiven for wondering whether they were dealing with the same subject. It is not just that 30 years of research have taken place; what has happened has been a 'quantum leap' in our ability to perform research on genes. This leap can traced back to the development of genetic modification and of a series of related techniques that followed in its wake. It has led, as one of us is very fond of saying to his students, to us knowing things about genes that we could not, in 1970, even have dreamed of knowing. This is what might be called the hidden application of genetic modification, at least insofar as it is hidden from the non-scientific public, who in general do not associate GM with media announcements about gene discoveries.

The reason for this research 'explosion' is that the techniques of GM enabled scientists for the first time to obtain large quantities (meaning large in biochemical terms, i.e. on a microgram scale) of genes and other DNA sequences. A spin-off from basic GM techniques, coupled with application of biochemical knowledge about DNA replication, led to the development of methods for determining the sequence of DNA and this, combined with studies of the expression of genes that had been transferred, led to understanding the relationships between sequence and function, especially in relation to gene control mechanisms. Taking plant genes as a particular example, these techniques, based around the use of GM micro-organisms to multiply copies of particular tracts of DNA ('molecular cloning') coupled with sequencing, were already being used extensively in research several years before genetic modification of plants themselves became possible. However, when it did become possible it added a further dimension to research, leading to extensive studies of gene activity and its control within GM plants.

Thus it became feasible to determine the DNA sequences of whole genomes (and, rather more slowly, to determine the function of all the genes and other sequences within the genomes, which is very much an ongoing process). The most famous example is of course the Human Genome Project (see Chapter 7), but several other genomes have been sequenced. These include 'model' organisms such as the mouse, the fruit fly, *Arabidopsis thaliana* (a model plant species), the bacterium *Escherichia coli*, several pathogens, including *Salmonella* species and *Plasmodium falciparum* (the protozoan that causes malaria) and its major mosquito vector, *Anopheles gambiae*, crop plants, including rice and more recently tomato and, for comparative purposes, mammals including chimpanzee and dog. The allocation of funds to genome research from government, industry and other sources across the developed world (many of these genome projects involve coordinated international collaboration) had an interesting knock-on effect. Because of the level of funding it became economically viable for companies specializing in scientific instruments and supplies to invest in extensive development in the technical support area. Thus, as one of us knows from hands-on experience in his laboratory, DNA sequencing has progressed from being a painstaking manual procedure to a fully automated process. Similarly, many of the procedures used in, for example, purifying DNA, isolating genes and gene cloning (via basic GM techniques) are now carried out with time-saving kits. All this raises interesting questions about allocation of resources, an issue with broad ethical implications and that has been the subject of much debate and discussion, particularly in the world of medicine. Extensive treatment of the topic lies outside the scope of this chapter but we invite our readers to consider the following question.

Question

In a world in which resources are limited, what would be your priorities in biological research?

The applications of this research on genomes have been far-reaching. Genomics, that is the study of the organization of genes within sets of chromosomes and the comparison of gene sequence and organization between different organisms, has become a new focus in molecular biology. We have a greater and ever-increasing knowledge about gene structure and function and of regulatory mechanisms. We understand more about evolution of genes, gene regulatory mechanisms and genomes. We can begin to think about new treatments for disease and better control of the agents of disease. We can use the information on plant genes and genomes to assist us in plant breeding and, for humans and indeed other mammals, we are reaching a greater understanding of genetic diseases and of the role of genes in pre-disposition to disease. All this can be traced back to the development of GM techniques in the early 1970s.

Dilemma

You are opposed to all forms of genetic modification, believing it to be an affront on nature. In the past you have demonstrated against GM crops. A farmer in your neighbourhood is planning to grow a new maize (corn) variety, which, according to an article in an agricultural magazine, will be more disease resistant and higher yielding than older varieties. It has not been bred by GM but by 'conventional' breeding. However, the breeding programme relied on an analysis of the maize genome that could only have been achieved by molecular cloning, sequencing and other techniques that are based on GM technology. What is your attitude to the new maize variety?

---

7 Human genes and the Human Genome Project

We are all mutants but some of us are more mutant than others

From *Mutants*, Armand Marie Leroi (2004)

We hold these truths to be self-evident: that all men are created equal

From *Declaration of Independence*, Thomas Jefferson (4 July, 1776)

7.1 Some history

Contrary to what one might read in the media, research on human genetics did not start in 1990 (when the Human Genome Project was initiated). Interest in human inheritance goes back a very long way and there were some remarkable insights into inheritance patterns centuries before there was any knowledge of genes. Plato, Hippocrates and Aristotle certainly made observations of the inheritance within families of particular characteristics. A few hundred years later, in AD200, the Jewish rabbi Judah the Patriarch deduced that a condition involving uncontrolled bleeding (leading to the death of baby boys after circumcision) was a familial trait. Of course, the condition he had observed was what we now know as X-linked haemophilia. This latter example introduces us to one of the main interests in human genetics, namely an interest in the inheritance of disease. There are about 4500 single-gene disorders (i.e. caused by mutations in single genes), most of which are rare. Amongst the best known of these are sickle-cell anaemia, cystic fibrosis, achondroplasia (skeletal dwarfism), and X-linked haemophilia. There are also conditions where there are interactions between genetic and environmental factors, and other conditions where a particular mutation may lead to a predisposition to disease.

Before the availability of genetic modification and associated techniques (including the direct detection of DNA sequences), study of any disease-associated gene was a frustrating business. For the most of these genes, studies

Introduction to Bioethics, by John Bryant, Linda Baggott la Velle and John Searle
Copyright © 2005 by John Wiley & Sons, Ltd.

of their inheritance depended on who had chosen to have children with whom. A couple concerned about whether they were at risk of having a child with a genetic disease could be given only a statistical probability, often not very accurate or specific, based on the pattern of inheritance of the condition in their families and knowledge of the population frequency of the particular mutation. There were just a few conditions where the estimates of genetic risk could be backed up by biochemical analysis of blood samples from the child after birth, for example in phenylketonuria. The ability to isolate individual genes (not necessarily an easy process, even with modern techniques, but certainly feasible) therefore received a very warm welcome that was echoed right across the world of molecular biology, whether particular laboratories were working on micro-organisms, plants, animals or human genetics.

7.2 Molecular genetics and the Human Genome Project

So, for a period of some 10–12 years after the development of gene isolation and DNA sequencing techniques, into the late 1980s, these techniques were applied in individual laboratories to the particular genes under investigation. This type of research, where sequencing is associated with a particular focussed project, is still very much part of molecular biology; indeed, one of us has direct experience of this. However, the past 15 years or so have also seen the initiation (and in several instances completion) of large, international, coordinated projects to sequence the genomes of particular organisms, some of which were listed in the previous chapter. The Human Genome Project (HGP) typifies this approach. Prior to the HGP, there was extensive activity in human genetic research in many laboratories all over the world. Of course, some of these laboratories collaborated with each other but in the main the research was not coordinated. Research focus was on genes of interest to particular biomedical scientists and some significant progress was made during this time, including the isolation of the cystic fibrosis gene and of several other genes involved in single-gene disorders.

Then, in 1988 a consortium of scientists in the USA persuaded Congress to fund a programme to sequence the entire human genome with the motivation of understanding not only heritable diseases but also those diseases based on molecular malfunctions in an individual, such as cancer. Interestingly, five per cent of the funding was set aside for a study of ethical and social implications of the project. In the main, the project was welcomed both inside and outside the biomedical community but there were some critics. A significant minority of scientists believed that this focus would 'skew' the balance of research so strongly towards genes that other areas of science would be starved of resources. This aspect of resource allocation was also raised both by clinicians and bioethicists, some of whom suggested that such large allocation to the project would divert attention from more important and widespread factors in disease, including poverty, malnutrition and poor living conditions. Further, there were some who believed that the potential for abuse of the knowledge was so great that such research should not be done. However, the climate in the USA was right for the establishment of a large prestigious project that would 'lead the world'. The project very much appealed to the American people, many of whom felt at the time that the USA had 'lost the space race' and thus needed to lead in something else. Is it very interesting that socio-political factors have such an impact on science progress. In the event, the USA did not in fact go it alone, because the project incorporated human gene analysis already in progress and stimulated further work (and funding for that work) in other countries, including the UK, Germany, France, Japan and Canada. Overall, therefore, about two-thirds of the project has been carried out in the USA, despite the impression still given in some American textbooks that the project has been entirely American (again giving food for thought on social factors in science).

The HGP was originally set to run from 1990 to 2005. Many scientists, including one of us, thought the timescale to be over-ambitious. However, such was the rapidity of the technical development (which we describe in the previous chapter) that the sequencing was finished two years early.[1] Scientifically, the data are fascinating, as three examples will illustrate.

- Any two individual humans differ in their DNA sequences on average by about 0.1 per cent (i.e. at about one base pair per thousand). This difference does not increase when we compare people of different 'races'; in other words those more obvious physical differences by which 'race' is defined are all contained within very small differences in DNA.

Question

Does this finding have any implication in relation to racialist attitudes?

- We only have about 25 000 genes, which is far fewer than the estimates of 'genetic functions', which run at about 80–100 000. It seems then that many of our genes (and indeed of the genes of all mammals) must be multi-functional.

---

[1] The exact date depends on what is meant by 'finished'. A 'first draft' was published over four years early and a 99.9 per cent accurate version of the sequence of 99 per cent of the 'gene-containing parts' of the genome was published in 2003.

- A direct comparison with the mouse genome has not only helped to unravel the function of some human genes that would have otherwise remained anonymous but also shown us that the human and mouse genomes differ by only a few hundred genes. All but about 300 of the mouse's 25 000 genes have direct equivalents in humans, and of the genes that humans possess, several hundred are not found in the mouse.

However, we need to recall that the original 'pitch' to obtain the support of Congress was that there would be significant medical benefits. Has this happened? Yes, to some extent it has in that some of the findings from the project have already been incorporated into clinical (mainly diagnostic) practice. We will discuss the use of human genetic data in Section 7.4, but at this point we need to take another brief look at history.

7.3 Some thoughts on eugenics

Over 2000 years ago, Plato suggested that human society might be improved by selective breeding and was thus the first to set down ideas of eugenics. The term eugenics may be approximately translated as 'well born' or 'good breeding', and it is linked in many people's mind with human genetics. How then did this situation arise?

The application of eugenics to human society was certainly something that Charles Darwin thought about, but it was his cousin Francis Galton who, in 1883, formally presented eugenic theory to British society. Galton's idea was that Darwin's evolutionary theory could be applied to humankind, that the quality of the human species could be improved if those with 'better' qualities produced more offspring than those with 'inferior' qualities. Those regarded as having inferior qualities included the 'criminal classes', the 'morally incompetent' and the 'feeble-minded' (the latter category described those whom we would today classify as having various grades of learning difficulties). These ideas were enthusiastically taken up by the Victorians and remained in vogue for several decades. Support for eugenic ideas declined in the UK from the mid-1940s (largely as a result of the second world war), but nevertheless there was still a small but clearly identifiable eugenic movement, albeit with little or no influence in wider society, up until the early 1960s.

Although eugenics in the 19th century was given to the world by a British scientist, it was in other countries that the concept was taken most seriously. For example, in the USA, a eugenic or 'stirpiculture' colony was set up in New York state in 1869 and an experimental breeding programme was initiated. Eugenic policies were incorporated into law in many states of the USA during the 1920s and 1930s, leading to the compulsory sterilization of the 'morally feeble' and of 'imbeciles'. Often, especially in the southern states, there was a strong racial element, with particular 'races' being regarded as inferior to others. Overall, it is estimated that in the 20 years leading up to the second world war around 40 000 people were sterilized in the USA for eugenic reasons. The practice declined during the 1940s and was ceased totally in the early 1950s.

Eugenic policies were also adopted in Canada and in several European countries, most notably in Germany, and it is the latter example that most people think of when eugenics is mentioned. Indeed, eugenic policies were taken to almost unbelievable extremes during the time of the Nazis, whose programme included sterilization (probably involving at least 400 000 people), experiments on humans, compulsory euthanasia and some enforced breeding experiments, as well as the extermination of millions of Jewish people in the name of racial purity. However, other countries also introduced compulsory sterilization on eugenic grounds, and in Canada, Sweden and Switzerland eugenic sterilizations continued until the 1960s.

Exercise

Evaluate the practice of compulsory sterilization on eugenic grounds, using in turn deontological, consequentialist, virtue-based and human-rights-based approaches.

7.4 Use of human genetic information

Introduction

Pause for thought

The philosopher George Santayana wrote 'Those who cannot remember the past are condemned to repeat it'. Is there any danger that use of human genetic information may lead to eugenic practices?

The HGP has provided new information about the involvement of genes in human disease and will continue to do so as the implications of the basic sequence information are worked out. The question thus arises as to how we will use the increasingly detailed and sophisticated knowledge and under-standing of human gene structure and function. We will look at this specifi-cally in relation to applications in medicine under the following headings:

- genetic diagnosis

- genetic screening

- genetic discrimination

- the burden of genetic knowledge.

Genetic diagnosis

One of the most obvious outcomes of the HGP is the increased availability of direct tests for the mutated genes that cause genetic conditions. In the mid-1980s only a very few direct diagnostic tests for 'disease genes' were avail-able. By contrast, at the time of writing this chapter in late 2004, the number had grown to about 340, although, because many of these tests relate to rare conditions, most clinical genetics centres will have only a fraction of them routinely available. Because these diagnostic tests rely on DNA, they may be applied at any stage of life (i.e. there is no need to wait for symptoms to develop), as follows:

- post-natal – after birth, in a baby, child or adult

- pre-natal – before birth but after the embryo has implanted into the wall of the uterus

- pre-implantation – before the embryo has implanted into the wall of the uterus

Post-natal diagnosis

The first point that must be made is that, although the new generation of gene tests makes use of DNA sequences, nevertheless, the implications of a positive test are not always clear. This is first because different people may express genes to different extents, particularly when the mutation concerned causes a strong disposition to disease rather than a total certainty. Second, there may be complex interactions between the gene and other factors such as environment, lifestyle and diet. Third, some genes are certainly multi-functional, so the effects of a mutation may be hard to predict. Fourth, the development of DNA tests is proceeding faster than the development of treatments, so even with a specific test result the person concerned may, in practical terms, be no better off. It is thus imperative that clinicians are very clear in the information they give about genetic testing.

However, this does not mean that genetic testing is worthless. In the UK, for example, all new-born babies are tested for phenylktonuria and congenital hypothyroidism because early detection allows the establishment of treatment and management programmes that will eliminate or at least alleviate the deleterious effects of these conditions. In some regions, the tests on new-borns also include thalassaemia and sickle-cell anaemia, especially in ethnic groups where these conditions are more common; in some regions cystic fibrosis is also included (see Section 7.4.3). These tests on new-born babies are pre-symptomatic, but genetic tests administered after symptoms have already developed can also be useful, confirming or refuting the initial diagnosis. Thus, parents who were unaware that they are carriers of the recessive condition cystic fibrosis find through genetic testing of their sickly baby that it has this condition (i.e. has two copies of the faulty gene). As yet, there is no treatment for cystic fibrosis, but even so the diagnosis will help the parents, and later the child him- or herself, to manage the symptoms. Diagnosis of particular conditions thus may lead to immediate help and perhaps even treatment for the condition. It may also spur the parents on to locate appropriate support groups and/or to obtain support through welfare and educational authorities. Similarly, testing of adults who have presented with particular symptoms may enable them to manage or even to obtain some treatment for their symptoms, to make appropriate lifestyle changes and so on. However, both for children and adults, a genetic test may reveal an especially distressing condition, perhaps of late onset, for which the prognosis is poor. In such cases, the knowledge itself may be difficult to bear. We discuss this more fully later in this section.

Pre-natal diagnosis

Pre-natal diagnostic tests for Down's syndrome and other *chromosomal* abnormalities have been available for over 30 years and some *genetic* tests for about 20 years. As indicated earlier, the number of gene tests available is now well over 300, but in general genetic tests are only carried out where the family history indicates that the foetus may be at risk of having a particular genetic disease, for example because the disease has appeared in previous generations or because both parents have discovered that they are carriers.[2] The tests are applied as early as possible in pregnancy (in practice from 11 to 12 weeks); in some cases, ultra-sound imaging may also be used in establishing a diagnosis. If the test is positive, termination of the pregnancy

---

[2] Not to be confused with chromosomal disorders: the main indicator for likelihood of Down's syndrome is the age of the mother.

(abortion) is usually offered. (In the UK, under the terms of the Abortion Act of 1967, the medical criteria for offering termination include 'a substantial risk that if the child were born it would suffer from physical or mental abnormalities as to be seriously handicapped'.)

Chapter 11 is devoted specifically to abortion, but here we need to deal with it in the context of pre-natal testing. Some prospective parents will have no hesitation in going for the abortion; others will feel more comfortable about a termination at 12 weeks rather than at 16 or 17 weeks into the pregnancy (as happened with earlier generations of pre-natal testing) and will also therefore go ahead with it. However, some will express concern that any termination, whether at 12 or at 17 weeks, destroys the life of a potential human being. For them, the decision is difficult to make and the potential severity of the condition may be a factor in their decision. Even so, there are some who, because of their total opposition to abortion, perhaps on religious grounds, will bring a foetus, even with a very severe genetic condition, to full term.

As hinted at briefly above, there is a tendency that the more severe the genetic disease that has been detected by pre-natal diagnosis, the more comfortable the prospective parents feel about termination, on the grounds that severe suffering is thereby prevented. Making a right judgement about a foetus who will be born to a life of severe disability and suffering and a possible early death is very difficult because different ethical principles, each valid and indeed praiseworthy on its own, come into conflict. The child may elicit from its parents and carers remarkable qualities of unselfishness and devotion. On the other hand, the child itself may suffer badly and caring for him or her may place huge stresses on individuals and on the parents' relationship that are unbearable. Perhaps sometimes it is legitimate to at least ask whether it would be better had the child not been born.

Questions

If abortions are offered in respect of severe genetic conditions, how severe is severe?

What level of disability and suffering should be dealt with in this way?

Does this practice narrow our concept of 'normality'?

Does it discriminate against disabled people? For example, if we abort a foetus because it has skeletal dwarfism (achondroplasia) what does that say about attitudes to people with achondroplasia?

What about eugenics?

Therefore, although the prevention of suffering is a worthy ideal under all ethical systems it may raise problems. For example, does the current practice of offering abortions in respect of particular genetic conditions start a slide down a slippery slope, as some have suggested? Whether or not one accepts the slippery slope argument, it can be seen that prospective parents may seek, or may be pressured into having, a wider range of genetic tests, with the possibility that abortions will be offered in respect of conditions that, hitherto, had not prevented the living of fulfilled and happy life. Some will be comfortable with these developments but others will be very concerned.

Pre-implantation genetic diagnosis

Techniques that enable the amplification of small amounts of DNA mean that it is possible to make diagnostic tests on the DNA of a single cell. This has been the basis of developing genetic tests with very early embryos created by *in vitro* fertilization, a procedure known as pre-implantation genetic diagnosis (PGD). A single cell is removed at the eight-cell stage (Figure 7.1) and the test is applied to the (amplified) DNA from that cell. Thus, if a couple is at known risk of having a child with a genetic disease they may opt to have children by *in vitro* fertilization (which in itself is a demanding and some-

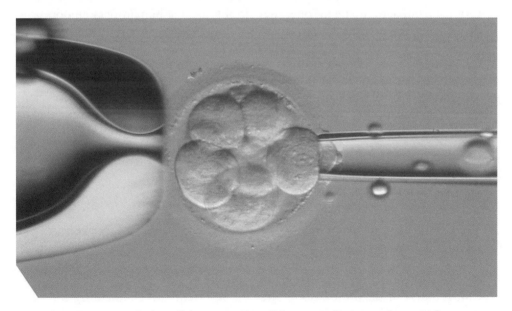

Figure 7.1 Removal of a cell from an eight-cell human embryo in order to perform pre-implantation genetic diagnosis. © Susan Pickering, King's College, London. Photograph reproduced by her kind permission

times traumatic procedure). Several embryos are produced. They are tested in the laboratory for the genetic disease at the eight-cell stage. Embryos free from the disease are implanted into the mother's uterus. Those with the disease are discarded. Thus, no pregnancies are established with the embryos that have the genetic condition and so the couple avoids the tricky decision about termination of pregnancy.

Exercise

Set out the arguments on both sides of this case, i.e. that pre-implantation genetic diagnosis and the discarding of the affected embryos is more ethically acceptable than aborting an established foetus that has a genetic condition and that this procedure is no different ethically from aborting a foetus with a genetic condition.

Whether or not PGD is regarded as ethical will depend on the view taken about the status of the early human embryo. Many people hold the view that although the early embryo has the potential to become a human person it is very far from actually being one: they do not ascribe human personhood to the early embryo. According to this view, the significant event at the start of life is the implantation of the embryo into the lining of the womb to establish a pregnancy. Only when this has happened can the embryo grow into a person. In the main, those holding such views regard PGD as an entirely acceptable way of preventing genetic disease from developing. However, those who believe that the very earliest embryo is a human person argue that discarding diseased embryos is ethically equivalent to aborting an affected foetus (see Chapter 11 for a fuller discussion of abortion).

Genetic screening

The term genetic screening refers to the practice of testing a large cohort of the population to see whether individuals have mutations that are likely to lead to specific diseases. We should also note that the word *screening* has strong implications of prevention (i.e. screening out). For example, in the United States in the 1980s, soon after the isolation and characterization of the gene that, when mutated, causes fragile-X syndrome, it was suggested that every foetus should be tested for this condition (followed by abortion of any foetuses with the mutation). Affected boys show varying degrees of learning difficulty and are often poorly coordinated physically. At the time, it cost $200 to administer the test, whereas, according to the company that wished to market the test, it would cost $2 million to provide lifetime care for a badly affected individual. However, the test itself does not indicate the likely severity of the condition. The arguments were purely economic and those who opposed the proposal, which included anti-abortion groups and a group representing some of the parents of boys with fragile X, did so on the grounds that the screening programme measured the value of lives, albeit lives affected by fragile-X syndrome, in purely monetary terms.

In the UK there has been a similar suggestion with respect to cystic fibrosis. The frequency of carriers of the mutated gene is between 1 in 25 and 1 in 22. At a carrier frequency of 1 in 25, the chances that a person and his or her sexual partner are both carries are 1 in 625; each of their offspring has a 1 in 4 risk of having both copies of the gene in mutated form and hence of having the disease. So, at a carrier frequency of 1 in 25, 1 in 2500 children will be born with cystic fibrosis. Of the genetic conditions that are regarded as diseases, this is by far the commonest in the UK; it has therefore been suggested that it is 'economically worthwhile' to undertake population screening for the mutated gene. There are however, some scientific complications: several different mutations are known and presumably screening would be focussed on the commoner ones, which make up about 80 per cent of the total. There is also disagreement about when to screen; detecting carrier status is the favoured option, but when should it be done in an individual's life and what does a carrier do with the information, for example in connection with relationships?

Thus far then, screening for cystic fibrosis has not been implemented nationally across in the UK. Nevertheless, it is a topic that is raised from time to time as being worth considering and in some regions a test for cystic fibrosis is one of those that are applied to new-born babies (in the 'heel-prick' or 'blood-spot' testing procedure). However, more recently an even more wide-reaching screening programme has been proposed: the UK Government has asked two of its advisory committees to consider the possibility of the genetic profiling, under the auspices of the National Health Service, of all new-born babies; this is referred to colloquially as the *bar-code baby* scenario. After lengthy consideration, both committees rejected the idea, at least for the time being. Nevertheless, it may be considered again in the future and thus we discuss it here. What this proposal entails is that DNA samples from new-borns would be screened so as to obtain a genetic profile (which might be confined to health-related information but could actually involve the whole genome). This genetic information would be part of the life-long electronic record kept for each person by the NHS. It is suggested that the information will help an individual to manage their health and lifestyle, will provide information on susceptibilities and will help clinicians to plan specific interventions and to come up with individually tailored treatments.

Question

Does this proposal raise any ethical issues?

Critics of this proposal are quick to point out that we currently lack the knowledge to predict what health implications a particular genetic change may have, let alone the effects of other factors such as environment and lifestyle on these possible outcomes. Even of those 300 or more tests that were mentioned earlier, not all provide an absolutely certain prediction of disease. On this basis, critics maintain that the idea is certainly premature and may indeed prove to be less useful than envisaged, even in the longer term, because of the uncertainties surrounding interactions between external factors and disease susceptibilities.

Two other issues have been raised by opponents of universal screening. First, there is the question of genetically testing babies without their consent (they are obviously much too young to give it). Supporters of genetic screening however point out that in law parents may give consent on behalf of a child who is too young to give it and that society in general accepts this. Thus babies are already tested soon after birth for up six genetic conditions, as described above. Second, it is suggested that the data will affect attitudes to people: that some people will be regarded as having less worth than others. In other words there is a danger of genetic discrimination, even as far as the creation of a genetic underclass, based on a DNA profile obtained soon after birth.

The thought-provoking science fiction film *Gattaca* (1998) envisaged a society in which *in vitro* fertilization was the norm. All *in vitro* embryos were genetically 'bar-coded' and parents selected only their 'best' embryos for implantation. Babies born by more conventional means (for example, in the film, as a result of unprotected sexual intercourse in the back of a car) were also genetically tested, and for both groups detailed predictions about likelihoods of contracting illnesses and about life-span were made on the basis of the 'bar-code.' Those born without the intervention of *in vitro* technology clearly had not been advantaged by pre-implantation genetic selection and were treated as less worthy beings, the 'Invalids', as opposed to the genetically selected 'Valids'. Will real life eventually imitate art?

We end this section with two real examples of genetic screening. The first concerns Tay-Sachs disease, which is common amongst Ashkenazi Jews. These are Jews from Eastern Europe, but many of their descendants live in the USA. Tay-Sachs is a neuro-degenerative disease causing, amongst other things, progressive loss of movement and an early death. The genetic condition is recessive: it takes two copies of the mutated gene to cause the disease. So, if two carriers each with a single mutated copy have children together, each child has a 1 in 4 risk of having Tay-Sachs disease. Wishing to avoid termination of pregnancy as a means of dealing with this, Rabbi Joseph Ekstein of New York set up a system for testing all young people for carrier status with respect to Tay-Sachs disease. The results are coded but not revealed to the young people who are tested. The test results are available when two young people begin to think of marrying each other and if they are both carriers they are advised not to marry. This is very hard for young people who are in love (or, in very traditional Jewish families, who have been brought together by the 'matchmaker'). Nevertheless, the programme is known within the community as *Dor Yeshorim*, a Hebrew phrase meaning 'the generation of the righteous', because it has reduced very dramatically the incidence of Tay-Sachs disease amongst Ashkenazi Jews in the USA and more recently in Israel itself, without the use of termination of pregnancy.

Exercise

This programme amongst Ashkenazi Jews has been called 'eugenics with a smiling face'. Discuss.

In Cyprus a similar approach has been taken in respect of thalassaemia, a disease in which the body fails to make haemoglobin, the oxygen-carrying protein of the blood. This is often a painful and crippling condition and sufferers from it require very frequent blood transfusions; even so they often die by the age of 20. In 1981, the Orthodox Church set up a programme in which it was insisted that all couples planning a church wedding should be tested for thalassaemia. If they are both carriers they are strongly advised not to marry. This again avoids difficult questions about termination of pregnancy but is equally tough on the two young people concerned. Interestingly, the church's approach to thalassaemia was in some ways a response to the programme set up in the 1979 by the Cyprus government. There is a government-funded testing programme for thalassaemia, very often applied to foetuses, accompanied by the ready availability of abortion (with which the Orthodox Church is unhappy). As with the Tay-Sachs programme, these measures in Cyprus have led to a significant fall in the number of babies born with the disease.

In both these examples, a particular population is at risk and measures are taken to avoid conception of an affected foetus or the birth of an affected baby. However, as noted already, the accepted courses of action may be difficult for the people involved. Are these examples where, instead of terminating relationships, *in vitro* fertilization and pre-implantation genetic diagnosis would now be appropriate? Perhaps once more the answer to this question will depend on views about the early human embryo.

Summary – genetic testing, screening and profiling

- *Post-natal diagnosis*: provides an explanation for suspected or observed symptoms; may enable informed treatment to be given and/or other forms of support to be obtained. Pre-symptomatic diagnosis of late-onset conditions may help in making lifestyle decisions. Burden of genetic knowledge may be difficult to bear.

  Can give information on carrier status. Programmes of testing for carrier status have led to reductions in the incidence of particular genetic diseases without use of abortion. Decisions about marriage between carriers may be difficult to make.

- *Pre-natal diagnosis*: generally used when family history indicates risk. Abortion is generally offered if test is positive. Decisions about abortions may be difficult, especially in considering the likely severity of the condition. In any case, some prospective parents will refuse abortion, whatever the condition.

- *Pre-implantation genetic diagnosis (PGD)*: involves genetic testing of embryos created by *in vitro* fertilization. Has been used by couples at risk of having a baby with a serious genetic disease. Only embryos free from disease are used to start a pregnancy. Many think that discarding the affected embryos is much more acceptable than abortion of an affected foetus. However, some think that discarding an embryo is ethically equivalent to abortion.

- *Genetic screening*: testing all members of a group or population, e.g. testing all foetuses for fragile-X syndrome or for cystic fibrosis. Supporters suggest that it will save costs of medical treatment by reducing the number of babies born with serious genetic disease. Opponents argue that it will lead to increased numbers of abortions and that it puts economics before people.

- *Genetic profiling*: applying a genetic test that covers the whole genome. It has been suggested that it should be applied to all new-born babies (the 'bar-code baby' scenario) so that detailed genetic information related to healthcare may be kept on record. This may help in lifestyle and health planning and in decisions about treatment. Opponents of the scheme are concerned that non-medical genetic information may be kept on record, thus abusing an individual's right to privacy, and that use of genetic profiles may lead to discrimination on genetic grounds.

The possibility of genetic discrimination

There is no doubt that the application of new genetic and genomic data to the understanding, at the molecular level, of human disease is beginning to be useful in real healthcare situations. An example of this is the rapidly growing number of genetic tests that are available, many of which are absolutely or strongly predictive of the likelihood of suffering from a genetic disease, including late-onset conditions such as Huntington's disease. However, the downside of this is that the increasing availability of genetic tests increases the possibility of discrimination on genetic grounds, as we have already noted in our discussion of 'bar-code babies'. One area in which discrimination has been shown to occur is the provision of insurance, both life insurance and health insurance. This is particularly important in countries where there is no state provision for healthcare. In one noteworthy case in the USA, health insurance has been refused for a child born with a genetic disease because the insurance company believe that the mother should have had an abortion after a diagnostic genetic test had given a positive result. Such cases seem straightforward but the insurance companies suggest that this is not so. Life insurance companies may argue, for example, that denying life insurance cover to (or greatly increasing the premiums of) people carrying a gene that is strongly predictive of a serious disease in later life is perfectly 'fair'. The companies also state that it is not fair for the financial burden of insuring the lives of such people to be borne partly by those who are free of such conditions. However, against this it may be retorted that until the availability of the genetic test the financial burden had indeed been spread amongst all the company's clients and so why should this not continue? Indeed, in general this is how insurance works: the many support the few. The debate thus continues.

In employment too, the situation is complex. Certain illnesses or physical impairments make it impossible for some people to hold particular jobs; this is accepted even in countries where there is legislation to support the rights of disabled people. There would thus be no pressure to employ someone who already had the symptoms of a genetic condition which made them unsuitable for the work in question. It is however much more difficult to decide the right course of action when a genetic test is strongly predictive of a serious late-onset condition. Furthermore, the situation is, month on month, becoming even more complex because we are achieving increased understanding that people vary considerably in their susceptibility to bacterial and viral infections, and are differently reactive to factors such as environmental carcinogens. Some of this variation is undoubtedly genetic in origin and it is very probable that tests for some of these susceptibilities will become available. What would the reaction be then to an employer checking on the vulnerability of a prospective employee to the common cold or to influenza, in an attempt to reduce absences caused by illness? Another scenario is that employers might favour people who are less likely to succumb to the effects of environmental toxins such as carcinogens, while paying less attention to chemical safety in the workplace.

The burden of genetic knowledge

In our teaching we often ask classes of students whether they would like to know if they carried a gene that either caused or gave a strong predisposition to genetic disease occurring later in life. There is always a majority who say 'Yes', but actually the evidence from wider society is that many people do not want to know, especially for those serious late-onset degenerative conditions such as Huntington's disease or familial Alzheimer's disease. The number of people requesting tests for these conditions is much less than would be predicted from the number of those likely to be at risk. For some, the knowledge that one is certain to suffer a serious and distressing condition is a burden too heavy to bear and thus ignorance is bliss. Such situations emphasize the importance of genetic counselling both in the phase of deciding whether to take the test and, if the test is taken, when the results are available.

*Case studies**

- Because of my family history I know that I am likely to be an unaffected carrier of a gene that causes a serious and so far untreatable condition. Do I request a test for that gene? If the test is positive do I tell my partner/ spouse?

- Knowledge of my family history informs me that I have a 50–50 chance of possessing a gene that around the age of 40 will cause a serious neuro-degenerative dis-ease for which there is no treatment. Do I want the test? If the test is positive, should I tell my partner, my children?

- I am currently healthy but I know that I have a gene that is very likely to cause serious health problems and possibly death in middle age. Who else should know?

In all these cases so far, we can add three more questions:

- First, should the knowledge go outside the family, for example to employers or to insurance companies: do they have the right to know?

- Second, how do I feel about the knowledge that I will or am very likely to suffer from a possibly lethal condi-tion? (The term 'pre-patient' has been used to describe people who are in this position.)

- Third, how do other family members regard such pre-patients?

The remaining questions concern the testing of children who are too young to give informed consent.

- My family history suggests that my child may possess a gene that will cause a serious late-onset condition.

- Should we have the child tested?

- If the test is positive, how do we then treat that child?

- Will the knowledge alter family dynamics?

- At what age should the child be told?

- What if, on reaching the age at which he/she can give consent, the child decides that he/she actually would have preferred not to know?

* These were published previously in *Life in Our Hands*, J. Bryant and J. Searle (2004) IVP, Leicester, and are based on study material developed by David Hardy, Bristol (see under *Myotonic dystrophy* at www.prs-ltsn.ac.uk/ethics/case_studies/index.html).

7.5 Genetic modification of humans – fact or fiction?

Introduction

Based on the techniques used for genetic modification of other mammals and on over 25 years experience in working with human embryos *in vitro*, it would be entirely feasible to attempt genetic modification of humans. This is not to say that the outcome of an individual genetic modification experiment could be predicted with any degree of accuracy. The variation in the level of expression of the foreign gene and its expression in subsequent generations would be subject to the same uncertainties that apply to other mammals. So, what is the current situation and what are the ethical issues that arise? We will discuss these questions under three headings:

- somatic cell gene therapy
- germ-line gene therapy
- genetic enhancement.

Somatic cell gene therapy

If, for example, a patient has an illness that leads to permanent kidney malfunction, the only effective cure is a transplant into the patient of a healthy kidney. It is in this light that we consider gene therapy. The rationale is simple enough. If a patient has a disease caused by a malfunctioning gene, then a 'gene transplant' may be a good way of curing the disease. But we immediately run into three problems. In dealing with the first we will assume that, as indeed is commonly the case, the condition has been diagnosed in a child. There is no way in which the functioning gene can be transplanted into all the cells of the child's body. So the gene is targeted to the particular cells which suffer from the effects of the gene malfunction. For cystic fibrosis, for example, the cells targeted are those that line the lungs, while for immunodeficiency diseases the bone marrow is the appropriate target. These cell targets are part of the already formed body (*soma*) of the patient, hence the term somatic cell gene therapy. A key feature of this is that the gene correction is limited to one generation only: the correctly functioning gene is not heritable. The second problem is that of actually delivering the gene. This is generally achieved by using a modified virus that will carry the gene into the target cell. Finally, there is problem of whether or not the gene actually works.

The motivation to bring benefit to seriously ill children has driven the development of gene therapy for a handful of diseases, including cystic fibrosis and severe combined immuno-deficiency disease (SCID). With cystic fibrosis there has been little success; gene function is at best only partially restored, and thus symptom relief is poor. Further, since the cells of the lung lining are constantly renewed, repeated treatments are necessary. The latter problem does not occur with SCID because the target cells are the self-renewing stem cells of the bone marrow (see Chapter 9); if the correctly functioning gene is inserted into the stem cell DNA, then it will be perpetuated throughout life. Indeed, there have been some spectacular successes in gene therapy for SCID: children who previously had been unable to fight off any infection were able to start to lead normal lives. However, several of these children have subsequently developed side-effects in the form of a leukaemia-like illness. It seems likely that the insertion of the functioning gene into a patient's chromosomes had activated an oncogene (a gene that, when it is switched on at the wrong time, causes cancer). This is another classic example of weighing potential harm against potential benefit.

Questions

When dealing with serious life-threatening conditions how far is it justified to use treatments that have great potential benefits but that are also very risky and that are essentially experimental?

Does it make any difference if the patients are children, unable themselves to give informed consent?

Germ-line gene therapy

Surveys of public opinion show that in general there is strong opposition to the idea of genetically modifying humans in a way that allows the new gene to be inherited from generation to generation. However, many people, including some professional ethicists, make an exception if the genetic modification is directed at correcting a genetic illness. The reasoning here is that eliminating a genetic condition in a heritable way, i.e. in the germ-line, would bring benefit to subsequent generations as well as to the initial recipient of the correctly functioning gene. But what is the reality?

It is entirely feasible to insert a new gene into a human egg immediately prior to or immediately after *in vitro* fertilization, and then to establish a pregnancy by placing the GM embryo in a woman's womb. However, as we have mentioned before, success rates both in terms of the number of live births and in the activity of the inserted gene are likely to be low. Nevertheless, as techniques for genetic modification of mammals improve, the possibility of success with human germ-line modification will increase, leading to pressure to use it as a therapeutic procedure.

However, even if germ-line therapy is adopted as an acceptable technique, it is unlikely to be have wide application. The probable scenario is that a couple with an absolute certainty of having a baby with a genetic condition, for example if both prospective parents are homozygous for a recessive, harmful mutation (i.e. they both have two copies of the faulty gene), will request germ-line therapy when they are planning to start a family. The couple would opt for *in vitro* fertilization and the correctly functioning gene would be inserted into several embryos that, prior to placing any in the womb, would be tested for the presence of the new gene. Such examples are rare; in most cases where couples are at risk of passing on a genetic condition, not all the offspring will be affected. In this latter instance a couple may go in for *in vitro* fertilization coupled with pre-implantation genetic diagnosis. Germ-line gene therapy would not be necessary.

Further, even though germ-line gene therapy would be needed only very rarely, some hold the view that genetic modification of a future human being should never be allowed. Indeed, in the UK, under the terms of the HFE Act, genetic modification of very early embryos is permitted in experiments aimed at understanding developmental processes; such embryos are destroyed 14 days after fertilization. However, attempting to establish a pregnancy with a genetically modified (GM) embryo is currently forbidden, even if the modification has been directed at eliminating a genetic disease.

Question

Are there specific ethical objections to germ-line gene therapy?

Two main lines of argument have been raised by those who oppose germ-line therapy. First, it is not yet clear, in GM of large mammals, whether or not it poses any risks for succeeding generations, even if it has been shown to be safe for the immediate recipient. Second, it has been suggested that germ-line therapy – and indeed any form of direct genetic intervention – goes too far in altering our biological nature; 'playing God' is a term that has been used in this context (although it is not entirely clear what is meant, it is often used to imply an intrinsic objection to germ-line genetic modification). Further, there is concern, even amongst some who do not oppose germ-line therapy itself, that it may open the way for other forms of genetic intervention such as genetic enhancement. This is of course a version of the slippery slope argument. Is it in any way justified?

Genetic enhancement and designer babies

The concerns of those who adopt the slippery slope argument centre first on the idea that the technical developments needed for germ-line therapy make it equally possible to apply them for non-therapeutic purposes such as genetic enhancement – using GM technology to improve in some way a human embryo. Second, it is argued that the general acceptance of germ-line therapy will make it socially and emotionally easier to accept non-therapeutic use of human GM. Third, they point out that the technique of PGD already allows the selection of or the rejection of particular genotypes, which although used for a number of therapeutic reasons could equally enable prospective parents to select for or against particular features according to their own wishes. This may be regarded as another form of genetic enhancement. In addition to these arguments, people who hold a very traditional ethical view of the early embryo, attributing to it the full status of human personhood, will object to any manipulation of embryos. We discuss different aspects of this ethical position more fully in Chapters 9–11. Here we return to the other arguments that deal with genetic enhancement.

First we must eliminate the more far-fetched possibilities that often feature in science fiction and even, sadly, in documentary programmes on TV. We are not talking about designing football (soccer) players to perform well in the English Premier League, nor about baseball pitchers whose performance will ensure that their team wins the World Series. Neither are we talking about ensuring that a child will turn out to be a great clarinet player or rock guitarist. While it is certainly true that many of the physical features that enable some-one to be, for example, an Olympic rowing champion are obviously genetic in origin, it is also clear that qualities such as sporting ability and musical and artistic talent are very complex, influenced by many genetic and non-genetic characters. So, while we may envisage one day 'designing' a long-legged person with an abundance of 'slow-twitch' muscle fibres, we cannot ensure that she will turn out like Paula Radcliffe. So what can we manipulate? We can manipulate, both by pre-implantation selection and by direct GM, characters for which direct involvement of a gene (or small number of genes) has been identified. In this sense then 'designer babies' are a real possibility.

Question

Are there ethical objections to genetic enhancement via pre-implantation selection and/or direct genetic modification?

For some, the answer is 'No'. For example, the British philosopher John Harris, based at Manchester University, has written

> *If it is not wrong to hope for a bouncing, brown-eyed, curly-haired and bonny baby, can it be wrong to ensure that one has just such a baby? If it would not be wrong of God or Nature to grant such a wish, can it be wrong to grant it to oneself.*[3]

Although on the surface this sounds very plausible there is no evidence that this view is held by the majority of prospective parents. Indeed, couples who choose to have children generally accept and love them as they come, whether boy or girl, blue-eyed or brown-eyed, blonde or brunette. Harris goes on to say that if it becomes possible in the future to provide the child with characteristics that give it a distinct advantage in life (that is to say a greater advantage than might be effected by hair or eye colour) then that too will be acceptable. In his view this is no different from paying for private education or for intensive sport or musical training. For Harris, all these activities rep-

---

[3] See Harris, J. (1998) *Genes, Clones and Immortality*. Oxford University Press, Oxford.

resent the parents' wishes to give their child the best in life; there is thus no ethical difference between genetic enhancement of the embryo and paying for one-to-one tuition on the trumpet.

We leave aside for the moment the topic of societal inequalities raised by Harris's comments in order to examine more fundamental issues. Those who oppose the use of GM techniques in genetic enhancement (and currently this is majority of those whose opinion has been sought[4]) have a range of reasons for doing so. At one end of the spectrum there are those who hold that any form of GM is intrinsically wrong (a view that was discussed in Chapter 5). Then there are those who hold that the human embryos are not to be exper- imented on or to be selected or rejected for any reason, because each one is a human person. However, mostly objections to human genetic enhancement are based on the view that to choose specific genetic features of a child (without of course any possibility of the child – embryo – giving consent) turns that child in a very obvious way into an object of its parents wishes. Some have gone as far as to say that this is a local form of eugenics; this may be rather strong but many writers agree that it amounts to 'commodification' of the child. In specifically ethical terms this does not conform to the virtue- ethics approach of dealing with others as we would have them deal with us, nor with Kant's (deontological) categorical imperative that no human should treat another as means to an end. For the present at least, these views prevail and a clear ethical line has been drawn between germ-line gene therapy and genetic enhancement.

However, there is a further complexity in this discussion, illustrated by the following question.

Question

Is it always possible to distinguish between therapy and enhancement?

The fact that we ask the question implies the answer – 'No'. There are several forms of medical intervention that certainly appear to be no more than enhancement but that for some may be therapeutic. Various aspects of cos- metic surgery fall into this category. Breast reduction may be undertaken to make a woman look better but it may also relieve painful side-effects of

---

[4] As seen in the UK from the results of social attitudes surveys.

having disproportionately large breasts. On the other hand, enlargement of small breasts, currently a popular form of cosmetic surgery in the USA and the UK,[5] is claimed to make women more confident in themselves and is therefore held to be psychologically or emotionally therapeutic. Other examples include leg-lengthening surgery – there was case recently in the UK of this procedure being paid for by the National Health Service so that a young woman would be tall enough to follow her chosen career – and the administration of growth hormone to children of short stature, even if their lack of height is not caused by hormone deficiency. In all these cases, the boundary between therapy and enhancement is very blurred and there will doubtless be instances in genetic modification where the distinction is equally difficult to make.

We must also ask whether it is likely to happen. In the UK the answer is at present very clear: genetic modification of embryos that will be used to establish a pregnancy is not permitted. However, it is probable that pressure will mount to allow germ-line gene therapy in the very limited range of cases that were described earlier. On the other hand, it seems very unlikely that the doors will be opened to germ-line genetic enhancement or to pre-implantation genetic selection for non-medical reasons, the views of writers such as John Harris notwithstanding.

The situation in the USA is somewhat different. Public opinion on genetic enhancement is probably overall more conservative than in the UK, but there is not a national authority such as the HFEA to regulate these activities, and so several American commentators have indicated that it is just a question of time and money before genetic enhancement is attempted in the USA. For example, Gregory Stock of the UCLA[6] writes, very similarly to John Harris,

> *If we could make our baby brighter, or healthier, or more attractive, or . . . otherwise gifted, or simply keep him or her from being overweight, why wouldn't we?*[7]

Stock takes his argument further, suggesting that the affluent will be able to afford to buy their children genetic advantages denied to the wider population. Again there are echoes of Harris, who believes that buying genetic enhancement is no different from buying educational advantage or extramural music lessons. The driving force thus becomes economic and it seems likely that in the USA attempts at genetic enhancement will occur in the relatively near future. This raises much wider issues, such as the inequalities in society and the way that resources are allocated, issues that will exercise many of our readers but that lie outside the scope of the present discussion.

---

[5] Presumably because of what, rightly or wrongly, are held to be the views of men on what makes a woman attractive.
[6] University of California, Los Angeles.
[7] See Stock, G. (2002) *Re-Designing Humans: Choosing our Children's Genes*. Profile, London.

Summary of Section 7.5

- Somatic cell gene therapy – the 'transplant' of a properly functioning gene into cells that are especially affected by a genetic illness – has been attempted for conditions such as cystic fibrosis and SCID

- It is still very much an experimental procedure, risky and with a low success rate

- Questions of risk versus benefit in experimental procedures are raised

- The same questions occur in relation to germ-line therapy – gene therapy in which the added gene is inherited by subsequent generations

- This is not a procedure that would be required very often

- It is currently not permitted in the UK under the terms of the HFE Act

- Nevertheless, there is general public support for development of germ-line therapy

- However, some have raised ethical objections to any form of genetic modification of the early embryo

- The possibility of germ-line therapy leads to consideration of genetic enhancement, *via* genetic modification of the embryo and/or by pre-implantation genetic selection

- There are some who are in favour of germ-line genetic enhancement, but mostly opinion is against it

- Those who oppose genetic enhancement suggest that it turns children into commodities to fulfil their parents wishes

- It is recognized that distinguishing between therapy and enhancement may be difficult

- In the USA, where the regulatory frameworks are very different from those in the UK, it is likely that market forces will lead to attempts at genetic enhancement

- Thus, the possibility of being able to create a 'designer baby' may depend on wealth

8 Genes – the wider issues

Increasing the amount we understand about genetics will affect all future generations. Our generation is charged with the task of setting the foundations ... and building the first few stages. It is important, therefore, that we think clearly and plan well. We need to encourage the scientists, technologists and financiers to step outside the tower[1] and look at what they are creating. We need to encourage the public and policy-makers to stop moaning about the technology and the problems of containing it, and take the effort required to understand enough about it to make enlightened decisions.

Let's step out of the tower, move away from the shadow and enjoy the light. Let's make use of genetics and not let it make use of us.

From *Babel's Shadow*, Pete Moore (2000)

8.1 Introduction

One of the most interesting features of the debate on GM crops in Europe has been the way in which issues not directly related to the technology itself have been brought into play. Some of the most vigorous opposition to these crops has been based on socio-economic arguments, mainly centred on inequalities in ownership of the technology. Indeed, the possibilities that GM technology may lead to further exploitation of the poor by the rich are, in the views of some, enough to make the technology irredeemable.[2] However, when these issues are examined in detail it becomes apparent that they are not exclusive to the applications of GM technology in agriculture. Nevertheless, the GM debate has been a useful vehicle for debating these issues, acting as a lightning conductor for attracting opposition that could equally be directed at several other aspects of genetic technology, at the pharmaceutical industry or indeed at several other aspects of modern developments in science and technology. Now, there may be some of our readers who are

---

[1] This refers to the Tower of Babel, as in the title of Pete Moore's book, *Babel's Shadow*.
[2] See, example, *Selling Suicide*, the 1999 report on GM crops by Christian Aid.

Introduction to Bioethics, by John Bryant, Linda Baggott la Velle and John Searle
Copyright © 2005 by John Wiley & Sons, Ltd.

content to leave it there, seeing socio-economic ethical issues as being outside their concern. We, on the other hand, suggest that scientists should be concerned with the way in which their discoveries are taken up. This is, as we argue more fully in Chapter 13, part of the social responsibility of the scientist. So, in the remainder of this chapter, we discuss three of these socio-economic issues arising from genetic research and GM technology, namely

- crop GM technology, world trade and global justice

- gene patenting

- genetic piracy.

8.2 Crop GM technology, world trade and global justice

Although there are many millions of hectares world-wide devoted to the growth of GM crops (as noted in Chapter 6), the majority of this land area is in developed countries or in rapidly developing countries, for example China. However, crop GM technology is often cited as being a key element in the fight against world food shortages, an important tool for the plant breeder (again as discussed in Chapter 6). Thus, the Director of the Rockefeller Foundation, Gordon Conway, whilst not regarding crop GM as a panacea for world food shortage, certainly sees it as an important tool in breeding programmes that can thus make a significant contribution in increasing food production. Indeed, he wants to see a second 'Green Revolution', a 'Doubly Green Revolution', a focussed world-wide effort utilizing the efforts of geneticists, plant breeders and agriculturalists to increase world food production, similar to the first Green Revolution of the 1970s.[3] Conway is especially concerned that increases in world population are likely, sometime in the first half of the 21st century, to outrun our capacity to produce food. However, those who endorse this view are reminded, by the opponents of GM technology, of the following.

- The success of the first Green Revolution was patchy. The new high-yielding cereal varieties mainly did well in parts of Asia and in South America.

- This enabled India, for example, to move from being dependent on rice imports to being a net exporter of rice.

---

[3] A concerted effort by plant breeders and agricultural scientists in the 1970s to produce high-yielding strains of cereal crops.

- However, high yields require high inputs and thus some less-developed countries became dependent on import of fertilizers from more developed countries.

- The Green Revolution was ineffective in Africa, mainly because of incompatibility with local agricultural practices, and failure to utilize local indigenous knowledge.[4] If there is to be a second Green Revolution it must therefore be more sensitive to local conditions.

- Further, even with current food production capacity exceeding global requirements, about one-sixth of the world's population is hungry, but this is not caused by inability to grow enough food.

- This hunger is mainly caused by poverty, especially in Africa.[5] Even in India, with its vastly increased productivity, many are too poor to be able to afford adequate food.

- Further, political factors may affect food production. For example, during the past 10 years, agricultural productivity in Zimbabwe has declined very markedly under the policies imposed by the president, Robert Mugabe.

These are all very relevant points in the argument, but nevertheless Gordon Conway and those who support his position argue strongly that we need to continue efforts to increase productivity in order to avoid actual global shortages occurring in the future. Such shortages will exacerbate severely the problems of the world's poorest people.

However, there is one further factor to be taken into account. The Green Revolution was largely based on research and development by government-funded laboratories and agencies, by charities and by international, non-profit organizations, which formed a diverse network with a single focus. By marked contrast, most of the crop GM research and development in the world's wealthier countries is carried out by large trans-national companies. Thus, the majority of the research and the resultant knowledge is in the hands of about six companies or conglomerates while government-funded and other non-commercial organizations currently play a relatively minor role in these developments.[6] Whatever we think about the profit motive, whether or not we are in general happy with the capitalist economic system, it is clear that

---

[4] This is discussed very fully in Chapter 11 of *Bioethics for Scientists*, eds Bryant, J., Baggott la Velle, L. and Searle, J. (2002).
[5] A situation portrayed poignantly by the rock band U2 in a recent song, *Crumbs from Your Table*, which includes the words 'Where you live should not decide Whether you live or whether you die'.
[6] This problem is not confined to GM technology. In 2004, 85 per cent of the world's tea (mainly grown in poorer countries) was traded by just three large trans-national companies.

there are problems in reconciling the need to make money with the application of a company's technology in the world's poorer countries. Indeed, the commercial practices of some trans-national companies certainly indicate that profit motive outweighs other concerns. Further, some economists have suggested that the way that world trade is organized, especially under the auspices of the World Trade Organisation, gives the wealthy countries of the world increased advantages over the poorer countries. However, other economists argue, equally strongly, that the WTO's operations will eventually lead to fairer trading condition. All these factors have led to the suggestion that at present crop GM technology is a powerful tool in the hands of the already economically powerful that may all too readily be used to exploit the poor and the weak. For some, the argument stops there.

Question

If you believe that crop GM technology can contribute to global food security, what measures would you take to ensure that the technology was available for application in the world's poorer countries?

There are voices saying, as we have indicated already, that this question is irrelevant. GM technology, it is said, is a 'high-tech fix', representing the unacceptable role of 'big business' in agriculture and food production and therefore not appropriate for the indigenous agriculture of less-developed countries.[7] It is certainly true that developments in plant breeding, whether or not including GM, should be relevant for local conditions and agricultural practices. Farmers who work hundreds or even thousands of hectares of land and who are used to buying seed anew each year have different needs and face different problems from farmers working small parcels of land and who keep back each year some seed for sowing in the next. The failure in much of Africa of the Green Revolution in general and of F_1 hybrids[8] (which do not breed 'true' from year to year) in particular is a testament to this. Further, there are certainly some small farmers in less developed countries who are opposed to using GM crops, possibly because of fears about being 'locked in' to an expensive dependence on wealthy commercial companies.

---

[7] This position is set out more extensively in Chapter 9 of *Bioethics for Scientists*, eds Bryant, J. Baggott la Velle, L. and Searle, J. (2002).
[8] First-generation hybrids between two elite strains of crop, which often outperform the two parental strains but which are very unpredictable in subsequent generations.

However, this is far from being the whole story. Several of the most rapidly developing countries, including India and China, have adopted appropriate GM varieties, as have cotton farmers in parts of South Africa, including KwaZulu-Natal. Both in China and in South Africa, farmers working small parcels of land are among those who, it is claimed, have benefited from growing GM crops. Indeed, based on the weekly electronic newsletter entitled *CropBiotech Update*, circulated by the International Service for the Acquisition of Agri-biotech Applications, SEAsia Center (ISAAA) and AgBiotechNet, agri-biotech applications for developing countries are developing fast (see knowledge.center@isaaa.org). Further, agricultural scientists and policy makers who are citizens of and based in a number of less developed countries have suggested that GM technology, appropriately applied, may be one factor in establishing local food security. However, can this be achieved without increasing dependency of the poor upon the rich or increasing exploitation of the poor by the rich? The answers to these questions involve, at one end of the scale, global economics. However, more local and situation-specific answers are also arising. Effective measures may include

- partnering arrangements between less-developed countries and major international agencies (such as the FAO) and/or other international non-profit organizations

- specific support, *via* international agencies, for appropriate research in those less-developed and developing countries that have the necessary research infra-structure

- recognition in national and international research programmes of what is relevant for particular places; in national research and development programmes this will lead to crops being bred for local conditions

- local individual arrangements that bypass or in other ways alleviate the problems relating to intellectual property (see the next section).

It is thus argued that GM technology may take its place, amongst other developments in plant breeding, within the range of measures needed to improve crop yield and quality in less-developed countries. According to supporters of this position, the development of 'Golden Rice'™ (a rice genetically modified to increase its vitamin A content), the growth and commercial success of insect-tolerant cotton in Kwa-Zulu and the large numbers of small farmers in China who grow GM crops all point to the success of such approaches. Opponents of GM crop technology remain unconvinced. But what of ourselves, the authors? This is one of those places in the book where we need to declare our hand: we suggest that, as a technique in plant breeding, genetic modification has the potential to be used in a non-exploitative manner in improving crop performance in less-developed countries. However, you, the reader, must make up your own mind.

Summary of Section 8.2

- World population is growing faster than agricultural production.

- It has been suggested that a second 'Green Revolution' is needed – a new effort to increase crop productivity.

- Supporters of crop GM techniques believe that the techniques have a significant role, as a tool in the hands of the plant breeder, in such a programme.

- However, if this is to be effective, some of the mistakes of the first Green Revolution, for example in Africa, must be avoided.

- There is also the problem that crop GM technology is largely in the ownership of a small number of multi-national companies.

- This, coupled with world trade patterns, raises the possibility of exploitation of poorer nations by the richer nations.

- Opponents of the technology point out that poverty is currently the major cause of hunger.

- They regard GM as an inappropriate 'high-tech fix'.

- Nevertheless, GM crops are being grown in several less-developed countries, where financial conditions are favourable or where partnering arrangements have been made.

8.3 Gene patenting

Gene patents in crop GM technology

Having just dealt with some of the problems associated with application of GM crops in less-developed countries, we move straight into a discussion of one of the more contentious issues that arise in this whole area. The key to the discussion is whether genes fulfil the criteria normally applied in consideration of whether a patent should be granted. So, to set the scene, here is the main question that needs to be answered.

Question

Is a gene an invention or a discovery?

The reason for asking this question is that in order to be the subject of a patent an object must be an invention and not a discovery or a pre-existing part of nature. Genes are clearly parts of nature; an individual gene, however ingenious the scientist has been in discovering it and characterizing it, is not an invention – end of story, or so one might think. However, it is not the end of the story. Genes, including crop genes, have been patented, most often in the USA but more recently within the patent jurisdictions of other countries, including the UK and other EU countries. How can this be? The essential argument made by those who support these patents is that there *is* an inventive step. The patent may be granted, it is stated, because the steps required to isolate the gene from the rest of the DNA and/or to make a copy of it from a messenger RNA population turn the gene sequence into an invention. Thus it is not the gene itself that is being patented but a copy made in the test tube. Opponents of patenting genes may well recognize the skill of the molecular biologist but they will add that it is a complete fudge to argue that patenting a gene copy is not the same as patenting the gene. They will also point out that defining genes as intellectual property in this way is a much more restrictive arrangement than the well established system of Plant Breeder's Rights or Plant Variety Rights. This results in yet more potential for the exploitation of the world's poor by rich commercial interests.

Supporters of gene patenting, however, having argued that the gene sequence is legitimate intellectual property, go on to state that this a logical extension of the internationally agreed arrangements for granting patents.[9] It

---

[9] Despite the fact that the EU patent jurisdictions did not initially follow the USA in granting patents on genes – they did not think that genes were patentable entities, but have since then changed their minds.

is argued that the particular way that GM technology has evolved means that this is the only way that companies can ensure an appropriate return on their research and development investment. As for the effect on less-developed countries, the example of the vitamin-A-enhanced Golden Rice™ is often cited. Although this was developed in non-profit laboratories, several patents stood in the way of its application. However, it proved possible to negotiate without cost 'freedom to operate' (FTO) agreements in all those instances where a patent would have otherwise proved restrictive. Of course there is no way of guaranteeing that such agreements may be reached in subsequent cases. Indeed, opponents of gene patenting point to the long battle in the world of pharmaceuticals before major companies gave up their intellectual property rights in order to allow the synthesis of generic drugs to deal with that other great scourge of Africa, AIDS.

Gene patents and medical genetics

As well as being a contentious issue in crop GM technology, gene patenting is equally so in the applications of human genetic information. The key question is the same – are genes discoveries or inventions? – and again the biotechnology companies have argued that making copies of genes allows them to be classed as inventions. There has been very widespread opposition to patenting human genes; some organizations, including, in the UK, the Nuffield Council on Bioethics, have opposed patenting of genes from any source. Gene patenting is also strongly opposed by the Human Genome Organisation[10] (HUGO): '. . . *the genome is the common heritage of humanity.*' Indeed, this stand against gene patenting taken by the non-profit organizations involved in the Human Genome Project was the cause of significant tension between them and a commercial organization, Celera Genomics. This company was not part of the public- and charity-funded HGP consortium but, having reviewed the commercial potential in the use of human gene sequences, had purchased 300 DNA sequencing machines and had sequenced at least part of most human genes by the time that the HGP consortium was ready to announce the first draft of the sequence. It had been the company's intention to patent these sequences but HUGO and the HGP consortium were determined that as many of the sequences as possible should be in the public domain and indeed had been placing each newly determined sequence in the public databases. The point was well made by Professor Bartha Knoppers, HUGO's chair of ethics, stating, in relation to the application of our knowledge of human genes, '*In the interests of human solidarity, we owe each other a share in common goods, such as health.*'

---

[10] HUGO is the umbrella organization that administers and coordinates the HGP. The HGP consortium is the group of laboratories in different countries involved in the research.

However, as we made clear in the previous chapter, human gene sequencing was under way before the HGP was initiated and even during the project significant activity in human gene sequencing took place outside the HGP consortium. Inevitably then, given both the commercial interest and the interpretation of the patenting criteria by the US patent office, human genes (at least 1000 as in early 2005) have been patented. There is clear evidence that, for genes that are already useful for genetic diagnosis and testing, patenting has affected the availability of the tests. Examples of this are the BRCA 1 and BRCA 2 genes, mutations of which give a very high lifetime probability of contracting breast and/or ovarian cancer. All the relevant data indicate that tests involving these sequences are more expensive than they would have been had the genes not been patented. This has implications for access, whether healthcare is provided through insurance (if costs increase, then premiums may have to follow) or as part of the social wage, as for example, in the UK's National Health Service (there may be questions of priorities in relation to spending a defined budget). Above all, it has implications for the less-developed countries of the world, especially sub-Saharan Africa. Average life expectancy in many African countries is less than 40 years; childhood dysenteric diseases and malaria are still major killers and HIV-AIDS is rife. If gene-based treatments do turn out to be useful in Africa, surely the need to pay increased costs due to patent protection would be impossible, as it proved with the anti-HIV drugs prior to the completion of negotiations on generic versions. Other cases where problems arise from gene patenting will surely follow and the arguments will go on. We therefore close this section with the following.

Exercise

Set out the arguments for gene patenting and against gene patenting, using any type of ethical system that you deem to be appropriate.

8.4 Genetic piracy

Piracy – the word conjures up a picture of a by-gone age: robbery at sea carried out by sailors brandishing cutlasses and wearing high boots and striped jerseys. It even has a slightly romantic image. However, it was not, and still is not, a romantic activity. Robbery at sea is still robbery. In those by-gone times pirates were very much feared, and even today in some parts of the world piracy is still a hazard, albeit not as frequent as in previous centuries. So what has piracy to do with genes? Can genes be the subject of a robbery at sea? To unpack this we need to note that that the term has gained other meanings since the 16th and 17th centuries. These meanings centre around using something without permission, such as running a radio station without proper authorization ('pirate radio'), infringement of copyright (as in the pirating of music CDs) or the infringement of another's business rights. Gene piracy embodies the concept of using genes (perhaps for commercial advantage) without permission.

Consider the following case.

Case Study

A man presents with symptoms that are shown to be caused by a rare form of cancer and as part of his treatment his spleen is removed. The pathology department at the hospital use it to establish a cultured cell line in order to study the rare cancer. The cell line performs so well that the scientists collaborate with a biotechnology company to patent it. They therefore start to earn royalties from other laboratories and organizations that wish to use the cell line.

On returning to the hospital at a later date, the patient is amazed to discover that his spleen cells have become a commercial entity, that the weight of cells derived from his spleen is now considerably greater than that of the original organ and further that those cells are distributed between several laboratories in different cities. All this has happened without any contact with the patient; his permission has not been sought. He has not even been informed of any of these developments.

Analyse the ethical issues arising from this case.

This may seem far fetched but it is loosely based on a real case in the USA. Is it genetic piracy? It certainly appears so on the surface. The patient's cells, for the sake of the lesion they exhibit, have certainly been used without the patient's permission and in a way that brought gain to the users. Let us attempt to dissect the case a little further. In terms of medical ethics, removal of the spleen was an act of doing good – beneficence. The slight inconvenience of living without a spleen was significantly less than the threat to the man's health had the spleen not been removed. Nevertheless, we suppose that the patient's personal autonomy had been respected in that he could have refused surgery had he wished. After surgery it was assumed that the patient had no more need for his spleen; indeed, it was for the sake of his health that it had been removed. Once outside his body, those involved deemed that he no longer had jurisdiction over it. In the real case, which involved a patient with hairy cell leukaemia, the patient sued the doctor and the university hospital involved but lost.[11] It was argued that, having given permission for the spleen to be removed, he no longer had any ownership rights to it. We may argue that the action of a virtuous person would have been to at least inform the patient as to what was intended, and more virtuous still to have asked permission. We may feel uncomfortable that an injustice has been done; we may also think that the patent on the cell line should not have been granted (but see Section 8.3, above). However, the law was not broken.

In the UK there is great sensitivity concerning the fate of organs removed for example, during autopsy examinations. This follows some high-profile cases concerning at least two major hospitals where organs removed from children who had died were kept by pathologists without seeking permission from parents. Under new and clearer legislation (The Human Tissue Act, 2004), if there is no pre-death consent of the deceased, next of kin's permission must be obtained to retain any organs from dead bodies (including organs needed for transplant) or to do any post-mortem research. Specific permission must also be obtained to use patients as subjects in research projects. However, as in the USA, it appears that once an organ has been removed *during surgery*, it no longer belongs to the patient, even if some like to keep their appendix or a diseased kidney in a jar in their office (and in any case, what exactly is meant by ownership of our bodies or their constituent organs is not very clear). Let us however imagine that permission was needed to use such an organ for research; would the 'donor' have any claim on income gained as a result of that research? Again the answer is 'No'. The situation would be similar to that of, for example, the live donor of a kidney: he or she has no call on any income that the recipient earns in the extra years of life gained. Anyone who donates a kidney makes a gift,[12] not an investment in the recipient.

In the case just discussed, things were not what they initially seemed. What about the following?

[11] John Moore vs The Regents of the University of California, 1990.
[12] We are aware that in some countries kidneys are offered for sale, but that is not relevant to the current discussion.

Case Study

A research team from a large trans-national pharmaceutical and agri-chemical company visits a small country in South America. They are interested in medicinal plants and focus on three species that grow in rain forest clearings and that are used in traditional medicine. On returning home with living plants and freeze-dried samples, they quickly establish that one of these species is a rich source of a compound that has great potential as an anti-inflammatory. The plant in question has an extra enzyme (encoded by a specific gene) that carries out the final step in the synthesis of the compound. The company initiates three lines of research: 1, investigating the possibility of chemical synthesis of the compound from its immediate precursor; 2, using cell and tissue cultures of the South American plant to see whether the compound can be synthesized in commercially viable amounts under controlled conditions; 3, in a long-term study, investigating the possibility of transferring the relevant gene, by GM techniques, into a crop plant that is easy to grow.

  What are the ethical issues in this case?

Here, the key questions again relate to ownership. First, in what sense can the indigenous people of the country be said to own the knowledge that certain plants help with pain relief? Is this 'intellectual property' in the sense that we normally understand it in the commercial intercourse of developed countries? The answer to this is almost certainly 'No'. Folklore does not sit comfortably with our systems for defining intellectual property and, in that sense, the scientists may argue that they were taking from no-one. Second, what of the plants themselves? Do wild plants belong to anyone? Certainly in many developed countries there are laws preventing removal of plants from privately owned land, but this was not the case in the present study. Further, the UK and certain other countries also have laws forbidding the removal of wild plants from their natural habitat, except under well defined circumstances. It is highly unlikely that the South American country in question had such laws. Presuming that the plants in question were not endangered species and that the returning scientists paid attention to the plant hygiene regulations in force in their own country, it is difficult to establish that they have taken anything that they should not have done. In some respects they were like those earlier generations of plant hunters who returned from distant shores with exotic plants that are now commonplace in our gardens, conservatories and greenhouses.

It thus appears that our scientists have done nothing illegal in bringing back these plants and in initiating a research and development programme that will lead to the registration of intellectual property in the form of patents and eventually to profits for the company, with no obligation to the country from which the plants were obtained. If this is piracy, who has been robbed; whose intellectual property has been used without permission? We could leave the case there, and if we did so we could cite the development by the major US pharmaceutical company Eli Lilley of the anti-cancer drugs vincristin and vinblastin, obtained from the Madagascar periwinkle, *Vinca rosea*. These have been both a medical and a commercial success but the people of Madagascar have not reaped any financial benefit from this.

However, we suspect that many of our readers feel uncomfortable about this and will at least have had the reaction that it is not fair. Surely, some will argue, the action of a virtuous person or even a virtuous organization would be to reward in some way the indigenous people on whose folk medicine the new drug is based, or if not the indigenous people, maybe the country at large could benefit. They may state this even more strongly when it realized that in the search for new plant-derived drugs, surveys that focus on plants used in traditional medicine have been much more effective in yielding interesting compounds than more random surveys. But, it remains clear that Eli Lilley did not break the law of any country in exploiting the Madagascan periwinkle.

However, the international community has moved to address this clear imbalance of power between the richer and the poorer nations. First, the 1992 Convention on Biological Diversity (often called the Rio Declaration) gave each sovereign state the rights over the biodiversity existing within that state. As we noted in Chapter 3, this includes the right to exploit commercially any living organism or any ecological community and some applications of that right may in fact have deleterious effects on biodiversity. In the case presented in the current study, the country in question would be able to force the company to enter into a specific agreement, perhaps allowing exploitation of any medicinal plants, in return for a generous share of any income that arises. This would avoid the type of situation that has arisen in Madagascar. Thus, Costa Rica, in Central America, has entered into an agreement with a transnational biotechnology company, enabling the company to exploit the gene pool of the country's rain forest under these terms. It is in the company's interest to protect their asset and thus to investigate the commercial potential of forests plants without destruction of this unique habitat. Overall, some commentators have suggested that agreements such as this might create a genuine commercial flow of money from the richer to some of the poorer nations. Second, the Rio Declaration recognizes the wealth of local knowledge on biodiversity held by indigenous people. As we have already noted, this folklore-based knowledge is not entirely compatible with more conventional approaches to intellectual property but the international community, acting through the World Intellectual Property Organisation,[13] is working to bring traditional knowledge under an extended intellectual property umbrella. This will be an attempt to ensure that indigenous peoples reap some reward, *via* internationally recognized mechanisms, if their knowledge is exploited commercially. Overall then, it appears that an imbalance of power is being corrected within this general area of exploiting exotic gene pools.

[13] See http://www.wipo.int/about-ip/en/studies/publications/genetic_resources.htm

9 Cloning and stem cells

I doubted at first whether I should attempt the creation of a being like myself, or one of simpler organization; but my imagination was too much exalted by my first success to permit me to doubt of my ability to give life to an animal as complex and wonderful as man.

From *Frankenstein*, Mary Shelley (1818)

9.1 Introduction

Fascination with human copies goes back a long way. In past ages, identical twins have been regarded as anything from sinister (especially if one of them was left-handed) to magical or even divine,[1] while stories of döppelgangers have appeared in a number of different cultures. The possibility of actually making human copies was typified in the 20th century by the 1978 film *The Boys from Brazil*, in which the notorious Dr Mengele was depicted as directing the cloning of a small army of neo-Nazis. 'Copying' specific people was the theme of Fay Weldon's 1989 novel, *The Cloning of Joanna May*. In the book, a man arranges that while his wife, Joanna May, is undergoing surgery, the surgeon will remove some cells from which the genetic material, DNA, may be extracted. This is then used to clone her, thus providing in the future 'new' versions of Joanna. The cloning theme has even appeared in cartoons, through the activities of the little boy Calvin in Bill Watterson's wonderful *Calvin and Hobbes* series. In 1990 the cartoon dwelt for several days on the theme of Calvin duplicating himself, urging the 'doubting Thomases' not to let ethics stand in the way of scientific progress whilst his toy tiger, Hobbes, expressed grave misgivings.[2] This fascination with cloning has a broad basis, which certainly includes our ideas of what makes us individuals and more recently on what role genes have in our development as persons. Neither is

---

[1] For example in an Iriquois creation myth, in which the world was made by a pair of twins, one left handed and one right handed
[2] See Watterson, B. (1991) *Scientific Progress goes 'Boink': a Calvin and Hobbes Collection*. Andrews McMeel, Kansas City, KS.

this fictional interest confined to writing about copying humans. The author Michael Crichton, probably aware of the well established procedure of cloning frogs (see below), based his 1980 novel *Jurassic Park* (made into a popular film by Steven Spielberg in 1993) around the theme of cloning dinosaurs from their DNA preserved in the bodies of blood-sucking insects trapped in amber.

Another common theme in the fictional presentation of cloning is that things can go badly wrong. In the *Boys from Brazil* the misuse of science, albeit highly fictional science, was at the centre of the plot. In the *Calvin and Hobbes* cartoons, chaos ensues as Calvin uses his 'duplicator' to make more and more copies of himself. And in *Jurassic Park*, the warnings of a more cautious scientist that the dinosaur cloners were on dangerous ground went unheeded, only for major problems to occur when some particularly fierce carnivorous dinosaurs escaped from their enclosure. Of course, such themes are not uncommon in science fiction; what might go wrong with the use of science makes for exciting plots. Nevertheless, as we mentioned briefly in Chapter 1, these ideas have entered the debates about the applications of science: it is often said that scientists do not know enough about the systems they are manipulating and would be unable to prevent either misuse or potentially disastrous accidents, but equally some of the scenarios envisaged belong in science fiction rather than in science.

9.2 Frogs and sheep

The examples mentioned above are all of course fictional, but did the authors and script-writers have any factual basis at all for the development of their plots? It has been clear for at least four decades that most cells in a fully developed multi-cellular organism retain all their DNA – their genetic material – even though only a particular subset of the genes is active in any one cell. During development from the one-celled embryo (zygote), there is a very complex programme of switching genes on and off. As investigations of gene activity gathered pace in the 1960s, one of the key questions in research was whether specialized cells retained the full genetic potential of the zygote, both in terms of the completeness of the information (is any lost during cell specialization?) and in terms of its activity (would all the genes still 'work' if they were placed in a situation where development would start again?).

In plants, it was demonstrated by cell and tissue culture that differentiated cells, subjected to appropriate treatment, could give rise to whole plants. Indeed, this plasticity of plant development is very helpful in the generation of whole transgenic plants from the initial transformed cells (as seen in Chapter 6). However, the organization and growth patterns of plants are very different from those of higher animals, and it took a different type of experiment to test the genetic potential of specialized animal cells. The question may be framed as 'can the DNA, the genetic material, of a specialized cell function as if it were back in the newly fertilized egg (known as the zygote, from a Greek word meaning 'coming together'), the very first embryonic cell?'. Framing the question this way indicates how the experiments were done. Frog eggs are large and thus the nucleus containing the DNA is relatively easy to remove. In this state, emptied of genetic information, the egg cell is incapable of any further development. However, if the egg nucleus is replaced by the nucleus from a specialized frog cell, then, under particular experimental conditions, the egg will start to divide and may go on to develop into a tadpole and then even into an adult frog. This process is known as nuclear transfer, and over the years its success rate with frogs has increased very significantly so that the procedure is now a routine part of particular research programmes on the regulation of gene activity during frog development. The procedure has been described in this way because the motivation for these experiments was based on this type of genetic research and the results clearly established, amongst other things, *that the genetic material of a specialized cell retains its full genetic potential and can be persuaded to 'start again'*. Of course, it is also true that the frog that results from the nuclear transfer is a genetic copy, a genetic clone, of the individual from which the donor nucleus was obtained. Interestingly, however, this was very much a secondary consideration when the experiments were first done in the 1960s,[3] and indeed remained so until the cloning of Dolly.

Question

Did the cloning of frogs by nuclear transfer raise any new ethical issues that should have been discussed when these techniques were being developed?

These nuclear transfer experiments in frogs were an important milestone in developmental biology. However, while experiments on frogs were becoming more and more sophisticated, all attempts at performing similar experiments with mammals failed. It looked very much as if the DNA of a specialized mammalian cell, although complete in terms of content, could not be reprogrammed to start again. Then, in February 1997, scientists at the Roslin Institute near Edinburgh in Scotland announced that a mammalian clone, Dolly the sheep, had been born some six months previously. For biologists,

[3] In the 1950s, frogs had been cloned from cells obtained from early embryos, but it was not until the 1960s that success was achieved with nuclei from cells of tadpoles and eventually of adults.

this was very exciting news: after over 30 years of trying it had been shown that the genetic material of an adult mammalian cell (for Dolly, a cell derived from the mammary gland of a six-year-old ewe) could be re-programmed to start again (albeit with great difficulty) – the changes were not after all irreversible. The major surprise was that this had been achieved not with the ubiquitous laboratory mouse, the subject of so much of the previous study, but with a large farm animal.

Questions

Does the cloning of a mammal raise issues that cloning of an amphibian does not?

Is animal welfare a greater concern with mammalian cloning than with amphibian cloning?

The science behind the cloning of Dolly is very important, and perhaps it was for this reason that the press officers of the journal *Nature*, in which the paper announcing the birth of Dolly was published, included details of the paper in their weekly pre-publication press release.[4] However, the cloning aspects clearly caught the imagination of the press and the editor of one major UK newspaper believed that the topic was so important that the paper broke *Nature's* date embargo and ran the story several days before the edition of *Nature* was published. This in itself raises an ethical issue that readers may care to think about. Media interest in the story was huge, at a level of intensity that biologists had never encountered before, and representatives of the press, TV and radio turned up at the Roslin Institute in large numbers. It was clear that the science of gene regulation was not the main topic on the minds of the reporters and news readers. Dolly was of course a genetic copy of the ewe from which the DNA had been obtained, i.e. was a clone, and this was the main focus of most of the media reports. Indeed, some of the media reports dwelt on the possibility of cloning humans, despite the clear statements from scientists at the Roslin Institute that this research was not intended as a step along that road (but the scientists also made it clear that Dolly was created as part of a programme to make genetic copies of valuable genetically modified ewes, and so it is probable that the cloning of sheep figured as strongly in their motivation as did solving problems of gene regu-

---

[4] We note that the paper's actual title was unlikely to generate a great level of media interest (*Viable offspring derived from fetal and adult mammalian cells*: Wilmut *et al.*,1997; see suggestions for further reading).

lation). In a further frenzy of reaction around the world, the Pope condemned cloning outright and the President of the USA (then Bill Clinton) requested that his Bioethics Advisory Committee should report on cloning as a matter of urgency, while the EU quickly enacted legislation to give all persons the right to their own genetic identity, in order to make illegal any attempt at reproductive cloning.

Questions

How open should scientists be about research that some may regard as controversial?

Should possible misuse of scientific results be a reason for not doing that particular research?

9.3 Genes and clones

Before discussing specifically the ethical issues, it is necessary to consider the relationship in humans between genes and individuality. Identical twins, with identical genetic material, developing in the same womb and growing up in the same environment, are not identical people. Anyone who wishes to clone a specific person will be disappointed. In Fay Weldon's novel, *The Cloning of Joanna May*, the husband looking for a new youthful version of his wife did not find her despite the strong physical resemblance of the younger women to their 'mother'. Similarly, cloning oneself, or even more tragically cloning a dying child, will not bring that person back again. Even at the physical level, the genetic clone may differ from the person from which the genes were obtained because of differing effects of environment, starting indeed with the uterine environment. Genetically identical twins may in some cases not look identical. It will also be impossible to mimic the factors that influence emotional and social development.

9.4 It's not natural – it should be banned!

The heading of this section reflects what the philosopher Mary Warnock has called the morality of the pub bore for whom 'It's not natural' apparently puts an end to the argument. However, as we stated in Chapters 2 and 6, natural versus unnatural is not a good basis for ethical classification, and in any case none of us, the pub bore included, could live with such a basis for our moral decision-making.

Question

Should there be a specific legal ban on human cloning?

Cloning in order to bring into the world a genetic copy of another person is a reproductive procedure and indeed is often called reproductive cloning in order to distinguish it from 'therapeutic cloning' (discussed later in the chapter). It would involve the collection of donated ova from women, the manipulation of these ova in order to remove the genetic material and replacing it with the donated genetic material, which would lead to the creation of embryos, albeit by very unconventional methods, and the insertion of some of those embryos into women's wombs. The result of this would be, if successful, the creation of a genetic twin of the person from whom the genetic material, the DNA, was obtained (the 'clone donor'). Thus in the UK cloning comes under the provisions of the Human Embryology and Fertilisation Act (1990; amended 2001) as administered by the HFE Authority. The Act and its interpretation by the HFEA are clear. Reproductive cloning is illegal in the UK. It is even illegal to split an embryo created by 'normal' *in vitro* fertilization in order to have identical twins (see Chapter 10).

The situation in the USA is somewhat different. It is widely stated that all forms of human cloning, including therapeutic cloning to make stem cells (see Section 9.6), are banned in the USA. However, further investigation shows this not to be so. Research on human cloning is banned in all federally funded laboratories (Prohibition of Human Cloning Act, 2001), but it is not banned in laboratories that receive their funding from private industry, from charity or indeed from any non-federal source. This does not mean that cloning research is proceeding apace in the USA, but it does mean that reproductive cloning could in theory happen there. Indeed, there have been several claims, all unsubstantiated, from an organization called Clonaid that several cloned babies have already been born in the USA. The refusal of Clonaid to produce DNA evidence throws this claim into great doubt, but what is clear is that there are medical scientists in the USA who, sooner or later, will be prepared to attempt reproductive human cloning.

At this stage we should ask whether there are in fact any intrinsic objections to human reproductive cloning, i.e. objections that would make us say that it is our duty to ban it. Presumably those scientists who are prepared to try it cannot identify any intrinsic objections. Neither can the Manchester philosopher John Harris, who believes that reproductive cloning should simply be evaluated as another reproductive technique. Indeed, cloning may be the only way for some couples, admittedly very few, to have a child that would be genetically related to at least one of them. The conditions that make this so are particular maternally inherited genetic diseases and certain forms of infertility. Further, some lesbian couples have suggested that having babies this way nicely by-passes the need for male gametes. Thus, it is argued, development of reproductive cloning would help couples who otherwise would remain childless. If this argument is accepted then some would push to the conclusion that reproductive cloning is acceptable for any couple.

However, even if one holds this view, it is clear that caution is still very necessary. The procedure still has a high failure rate at every stage, from nuclear transfer into the egg cell right the way through to birth. The excellent and informative web site of the Roslin Insitute[5] (where Dolly was born) shows that up to July 2002, six years after the birth of Dolly, success rates were still very low indeed, across a range of mammalian species. Furthermore, there has been no breakthrough on this since these data were posted. Attempting to clone humans at this stage in the development of the procedure would be to treat humans, and women especially, as experimental objects. In particular, any woman who becomes pregnant as a result of implantation of a cloned embryo carries a significant risk of experiencing the spontaneous abortion of a malformed embryo or perhaps worse of bringing to term a seriously malformed embryo. Many people, even those who do not have any intrinsic objections to reproductive cloning, would find it unacceptable to use women in this way, effectively as experimental material. Indeed, scientists at the Roslin Institute suggest that, notwithstanding any other arguments, these grounds alone are enough to prevent attempts at human cloning. Further, in terms of conventional medical ethics the very high risks of this procedure are not at present justified by the possible benefits.

Question

If the risks can be reduced to acceptable levels, is the treatment of fertility problems an acceptable application of cloning?

The objections set out above are essentially based on risk and in particular on the very high risks of failure, failure of a type that may prove very traumatic. No specific intrinsic objections are raised and thus, some would argue, the way is left open for reproductive cloning, should the risks become low enough to be acceptable at least in the context of offering help to certain infertile couples. Indeed, this type of argument has often been presented in

---

[5] www.roslin.ac.uk/public/cloning.html

the media. However, there are other risks that are not reduced by improvements in the cloning procedure *per se* and these are the risks to the clone him- or herself. Unlike *in vitro* fertilization, cloning by-passes the coming together of gametes (eggs and sperm) of different genetic make-up that sets up a new genetic mix (which is one of the functions for which sexual reproduction is believed to have evolved). Routine *in vitro* fertilization techniques, although they separate the act of sexual intercourse from the process of procreation, preserve this coming together of the genetic material from the two parents. Again it is emphasized that naturalness or unnaturalness are not in themselves strong factors in the ethical argument, but this very marked biological difference between cloning and sexual reproduction may be a factor in discussion. Further, the DNA that is used to support the development of the egg is the result of a long biological history and often needs unusual treatment in order to reverse its developmental state. It is not at all clear that the risks of these processes are acceptable or ever will be. Has the DNA accumulated a lifetime's worth of unrepaired damage? If so, will the new embryonic environment enable the DNA to be repaired? At present there are no clear answers to these questions. Dolly, the first mammal cloned from an adult cell, suffered early onset of degenerative disease, as if she were as old as her clone donor, but it is not clear whether this is a widespread problem. Human clones are thus exposed to unknown and unquantifiable health risks.

For further consideration of the ethics of cloning it is helpful think of the reasons (other than helping couples to overcome fertility problems) there may be for cloning a human. As mentioned already, some of the earlier enthusiastic and positive responses to the possibility of cloning humans were based on the wish to recreate a specific person such as a loved one. The mistake inherent in this idea has already been dealt with: a clone will reproduce a genotype (i.e. a specific set of genes), with no guarantee of how that clone will turn out as a person. Our personhood relies on very much more than our genes. We cannot manufacture a particular person by copying a particular genotype. However, because the genetic material used to create a cloned embryo is taken from a cell in an adult, it would be possible to have a good idea of how the genotype will be translated into the physical phenotype, in other words, how the clone will 'turn out' physically. Even this aspect of development is subject to environmental influences, again including the environment in the womb, but nevertheless, the likelihood of physical resemblance to the DNA donor might provide enough incentive for some people to try (as in the fictional situation presented in Fay Weldon's novel). Having said that much, it also must be emphasized here that mammalian cloning is not mass technology; furthermore, based on over 40 years experience of cloning frogs and nearly 25 years of experience of human *in vitro* fertilization, is very unlikely to become so. The projected scenario is not one of creating armies of drones to carry out menial jobs, nor of creating football teams. The film *The Boys from Brazil* often crops up in discussions of cloning but the plot, although good fiction, is very far fetched. Nevertheless, it is clearly possible to think of reproducing a specific genotype with characteristics desired by those attempting to bring a cloned child into the world.

Question

Can the selection of a specific genotype for cloning be in any way regarded as a eugenic activity?
(See Chapter 7.)

This possibility of cloning to copy a specific genetic make-up raises a number of questions, one of which has been posed immediately above. Another question is whether the procedure makes the cloned child a commodity in that there is the attempt to fulfil the specific wishes of other people. Whatever one thinks about treating the rest of the living world in this way (for example in the use of farm animals), treating another human as a commodity, i.e. in a specifically instrumental way, again contravenes any ethical code based on the autonomy, dignity and worth of the individual. However, it is also true that people have children for a variety of reasons, some of which look as if the child is a commodity (as discussed more fully in Chapters 7 and 10). This issue is not therefore specific to cloning, but has more general applications to the reasons for choosing to have children and to ways in which humans treat each other in general.

This leads us on to think about the emotional/mental health of any child who is born as a result of these cloning procedures. Will their very unconventional origin have any psychological effects? We simply do not know, although we can say that it is likely to vary from person to person. For people produced by cloning, the genetic mother and father are one generation further back than in normal sexual union; the clone did not arise from a conventional or even an *in vitro* fertilization and did not therefore have a mother and father in the biological sense, at least in the way in which we normally use those terms. Knowledge of this may well be disturbing. Further, the clone will have been created specifically to fulfil the wishes of others, leading to social and emotional pressure that some will find hard to deal with. However, the latter point is not specific to cloning. Many parents place expectations on their children and may make arrangements on their children's behalf to push them towards fulfilling those expectations, even to the extent of living emotionally through their children. While some children cope with this with no apparent difficulty, others rebel against fulfilling the wishes of their parents, while yet others may believe themselves to be failures or even experience feelings of rejection if they do not (or in their own minds do not appear to) live up to their parents' expectations. Would a person feel any greater pressure if those expectations were based on the fact that they were genetic copies of another person? Again we do not know.

9.5 The ethics of human cloning – an overview

In the previous section we have raised a number of problems, risks and unresolved questions relating to human cloning. The key points may be summarized as follows.

- It is a very uncertain procedure with a very low success rate

- Until this is resolved, women who take part in cloning procedures would be subjects of highly experimental and very controversial procedures

- There are unresolved questions about the health of cloned animals; these would also apply to humans

- The risks far outweigh any supposed benefits

- Except in instances where cloning is used as a last-resort procedure in fertility treatment, there will be questions about the motivation of those who wish to bring clones into the world

- The view that cloning can recreate a specific person is mistaken

- There may be emotional problems for cloned children, who will be under pressure to live up to particular expectations

- The unusual origins and the uncanny resemblance to the clone donor may be emotionally and/or psychologically harmful

In summary then, it is clear that from the start of the cloning procedure through to the life of the cloned person themselves there are several serious unresolved (possibly unresolvable) problems and risks. Although these may not be specific to cloning *per se* they are enough for many people reason to support a ban on cloning. However, as pointed out earlier in the chapter, there are some who argue that there are no intrinsic ethical objections to human cloning. Their position on the problems summarized above is that to ban cloning for these reasons would be to do so on consequentialist grounds.

In other words, it would be concerns about the consequences that might lead to a prohibition. Of course, some of the problematic consequences might melt away as the biomedical community improved the procedures used in cloning. However, there are other views in this debate.

For Roman Catholics and for some other Christians, cloning is forbidden because it is one of several procedures that separates sexual activity from procreation (a position that we also encounter in Chapter 10). This is a deontological viewpoint based on what are perceived to be the teachings of the Bible as interpreted by the Church. Moving into the cloning methods themselves, it is likely that in any cloning procedure attempts will be made to create several embryos and of those that reach the blastocyst[6] stage some will be rejected. This will raise objections from anyone who holds a traditional ethical view of the early embryo, however it comes into being; for example, some believe that the very early embryo, even before implantation in the womb, is a human person. To reject and destroy an embryo is, in this view, to reject and destroy a person. We discuss this view more fully in Section 9.6.2 of this chapter).

Thinking now of more widely held ethical positions, we have suggested that currently any woman undertaking to carry a cloned embryo and any cloned person who is born would firstly be 'experimental material' and secondly would be exposed to several types of risk, risks that could not be evaluated or quantified until several such 'experiments' had run their course. In Kantian ethics, one of the categorical principles is that one human does not use another as a means to an end, and this deontological position is obviously applicable to cloning. Further, this is similar to the virtue-ethics principles of treating others as we would be treated ourselves that are embedded in the practice of several religions, and also in humanism and in the Victorian 'golden rule'. Both these sets of principles (Kantian categorical principles and virtue ethics) also come into play in dealing with motivations for making a specific genetic copy of any person.

Two other ethical positions must be mentioned. The first of these is moral repugnance, a widespread feeling, perhaps based partly on the 'yuk factor', that this is wrong. We have argued elsewhere in this book that ethics based purely on the yuk factor does not work, but in some instances repugnance goes deeper than the yuk factor. Thus, ethicists in both the United Kingdom and the USA have suggested that repugnance about human cloning reflects a position, based in our common humanity (as emphasized by humanism and by most religions), that to try to make another copy of another person is simply wrong. The second position is based on human rights. Thus the EU now has legislation that asserts the right of every person to their own genetic identity. One wonders whether, prior to enacting this legislation in the months

---

[6] This developmental stage occurs a few days after fertilization, shortly before implantation into the lining of the womb is initiated. It is described more fully in Section 9.6 of this chapter.

following the birth of Dolly, there had been enough thinking about, for example, identical twins, or about those people who receive many copies of another's genome in the form of an organ transplant. Nevertheless, the message is clear: according to this legislation, cloning is not compatible with the human rights of an individual.

Exercise

The year is 2020. Legislation has been made to allow the use of human cloning in medical treatments. In effect, this was to regularize the use of cloned early embryos to make stem cells, something that had been allowed by the HFEA since 2004, but that, some argued, had broken the law that bans human cloning. The new law thus ensured that cloned embryos could be used, legally, as sources of stem cells.

Furthermore, cloning procedures have improved dramatically. The success rate in experimental animals is nearly as high as that achieved by in vitro *fertilization. One of the keys to this success is the careful sorting of the early embryos after a few days' growth so that only those in best health are placed in the womb. Furthermore, experiments with chimpanzees and monkeys have led to the problems previously occurring in primate cloning being largely overcome. It had been suggested that a human cloning procedure would have about a 10% chance of success, based on work with chimpanzees.*

Nicole and Matthew are a couple in their early 30s. Matthew does not produce effective sperm and so they have been unable to have children. They are unwilling to use donor insemination because of the involvement of, as they put it, 'someone else's genes.' However, they argue that by using cloning techniques, they could have children of their own without any need for a third party. Nicole can provide the eggs from which the DNA is removed, to be replaced in one instance with the nucleus from one of her cells and in the second instance with the nucleus from one of Matthew's cells. Thus they would hope to have a girl who was a genetic copy of Nicole and a boy who was a genetic copy of Matthew. They argue that allowing human cloning in medical treatments opens the way to its use in helping infertile couples such as themselves and they have applied to the HFEA to allow them to do this.

Set out the views that you think are likely to be presented and debated at the next meeting of the HFEA when the application from Nicole and Matthew is discussed.

9.6 Unlocking the genetic potential of stem cells

Embryonic stem cells

At the time of writing this chapter, one of the more controversial developments in biomedical science is research on human embryonic stem cells. The research is aimed at making it possible to use particular cells in the early human embryo as sources of 'spare parts' for tissue and organ repair. To understand this, it is necessary to describe the early stages of mammalian embryonic development, illustrated diagramatically in Figure 9.1.

As already indicated, the zygote, the fertilized egg, contains all the genetic information necessary for the complete development of the adult mammal. The zygote is thus said to be totipotent. This totipotent state is retained by all the embryo cells through the first few rounds of cell division until the blastocyst stage is reached. As shown in Figure 9.1, the embryo is at this stage a hollow ball in which a dense group of cells, the inner cell mass (blastocoel) hangs from the outer cell layer. During normal development, the inner cell mass grows out through the outer cell layer ('hatches') and begins to attach to the lining of the uterus, thus establishing a pregnancy. These cells of the inner cell mass go on to develop into the foetus. Meanwhile, the cells of the outer layer of the blastocyst give rise the placenta. Thus, at the blastocyst stage, the outer cell layer has lost the potential to develop into a fully formed mammal and the inner cell mass has lost the potential to form the placenta but it does of course retain its much wider potential of being able to grow into the whole living mammal. At this stage, therefore, the cells of the inner cell mass are stem cells, cells that have the developmental potential to form many different types of cell and, in this particular instance of embryonic stem cells, all the different types of cell that occur in the mammalian body. In *genetic* terms these cells are totipotent but because at this stage of embryonic growth they have lost the developmental capacity to form the placenta, they

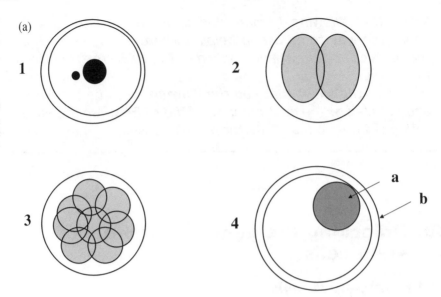

Figure 9.1 (a) Diagrams of the early development of the human embryo. Reproduced by permission from *Life in Our Hands*, Bryant, J and Searle, J., Inter-Varsity Press, Leicester (2004).

1 Shortly after the penetration of the egg by a single sperm cell, the two nuclei, each containing one set of genes, are clearly visible. Note that although the two nuclei contain equivalent amounts of DNA, the egg cell nucleus is much bigger than the sperm cell nucleus. The complete fusion of the two nuclei and the activation of the fertilized egg to divide will take about ten more hours.

2 The first cell division has taken place (note that the nuclei are not shown in diagrams 2–4).

3 After two more rounds of cell division the embryo is at the uncompacted eight-cell stage. The embryo is still 'indeterminate' and it is thus possible to remove one cell for genetic testing (as described in Chapter 7).

4 The blastocyst stage. The embryo now consists of an inner mass of cells (a), which will, if the embryo implants, become the embryo proper, and an outer layer of cells (b), from which the placenta will be derived if a pregnancy is established (see Chapter 11). Stem cell cultures may be established from the inner cell mass (blastocoel).

(b)

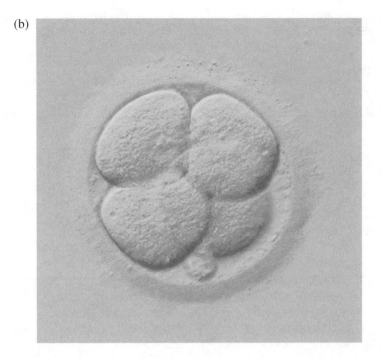

(c)

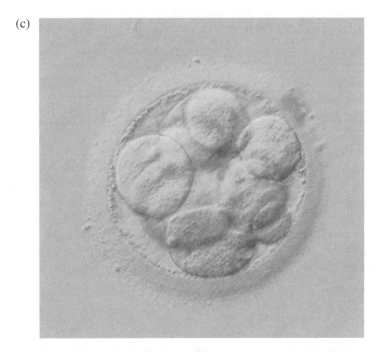

Figure 9.1   (b) Photomicrograph of a four-cell human embryo. ©Susan Pickering, King's College, London. Photograph reproduced by her kind permission. (c) Photomicrograph of an eight-cell human embryo. ©Susan Pickering, King's College, London. Photograph reproduced by her kind permission.

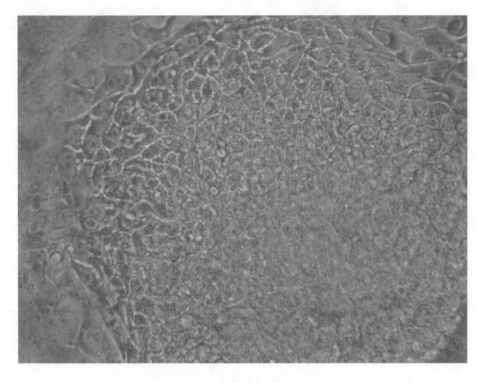

Figure 9.2 Photomicrograph of a culture of human embryonic stem cells. ©Susan Pickering, King's College, London. Photograph reproduced by her kind permission.

are not *developmentally* totipotent.[7] They are thus described as pluripotent stem cells.

Developmental biologists have had some success in persuading mouse embryonic stem cells to grow into particular cell types in the laboratory. This led the UK's HFEA to grant a very limited number of licences to particular laboratories for the culture of human embryonic stem cells, derived from the blastocyst (see Figure 9.2). As with the mouse, it has proved possible to induce the formation of a range of specialist types of cell from these human embryonic stem cells. This reinforces the idea that embryonic stem cells may in the future be used to repair tissues and organs damaged by disease or accident.

---

[7] Although readers will be quick to realize that developmental totipotency might be restored by using such cells as nuclear donors in cloning experiments.

Exercise

Outline the key ethical arguments that are likely to be raised in support of and in opposition to this research.

The terms of the original HFE Act in the UK allowed the creation of embryos *in vitro* not only to enable sub-fertile couples to have babies but also for specific research projects relating to human reproduction (see Chapter 10). However, in practice such research is mostly carried out with 'spare' embryos from IVF rather than with embryos created specifically for research, the latter amounting to a fraction of one per cent of the total. Nevertheless, following the recommendations of a committee chaired by the UK's chief medical officer, Professor Liam Donaldson, the HFE Act was amended in 2001 to extend the research use of *in vitro* embryos to include research on embryonic stem cells. However, although this opened the door to the creation *in vitro* of embryos specifically in order to generate embryonic stem cells, in practice nearly all the research on human embryonic stem cells is carried out with spare embryos from IVF treatments.

Embryonic stem cells and the ethical status of the early human embryo

The ethical discussions about these developments in the UK continue and indeed are likely to do so. First, there are those who view the earliest human embryo, the one-celled zygote, as a human person. The following points are often made in support of this position.

- Each zygote has a unique human genotype that has never existed before and will never exist again (unless the embryo splits to form identical twins)

- Given the right conditions in the womb the embryo will develop into a foetus and thence into a child.

On these grounds there can be no such thing as a spare embryo; it would like saying that a fully formed human was a spare person. In this view, IVF treatments should only involve creating one embryo at a time, with each embryo thus created being given a chance to establish a pregnancy by implantation in the womb. Further, the deliberate diverting of embryonic development to establish a stem cell culture is regarded as destroying a human person (and some would go as far as defining this act as murder). According to this view, it matters not whether the embryos are 'spare' ones from IVF treatment or have been created specifically for establishing a stem cell culture; their use in this way is wrong.[8] It will be obvious to our readers that this is a deonto-logical position.

On the other hand, the majority view in the UK does not bestow human personhood on the early human embryo. Thus it is pointed out that in nature up to 80 per cent of fertilized eggs/very early embryos do not implant into the lining of the womb and thus do not establish a pregnancy. If all these early embryos are lost then, the argument goes, it is difficult to regard each one as a human person even though each one has a unique set of genes. They also point out the following features of early development.

- It is not until several rounds of cell division have occurred that the allo-cation of specific cell lineages to placenta and to embryo is made

- Even after this, the embryo itself may split to form identical twins, suggesting that the early embryo cannot yet be regarded as a human individual.

- Studies of genetic mosaicism suggest that, as has been observed in other mammals, two very early embryos may on rare occasions merge to form one that develops normally.

On these grounds, the use of early embryos to create stem cell lines does not mean ending the life of another human. Further, the use of spare embryos may be regarded as an ethical good: these spare embryos, unless the 'parents' had given permission for research use, would, after several years of deep-frozen storage, be discarded. Their use for stem cell research can instead bring major benefits to existing humans and thus to society at large. This is a view strongly espoused by the Australian philosopher and bioethicist Peter Singer (now based at Princeton University, USA). Readers will recognize this as a consequentialist, and more specifically a utilitarian, argument.

Question

Is there an intermediate position in this debate?

---

[8] See for example http://www.lifeuk.org/news/news.php?subaction=showfull&id=1039178801& archive=&start_from=&ucat=& where very early embryos are described as 'tiny human beings'.

Whilst the previous paragraphs have set out the two main positions in the debate about stem cells, there are others who ask whether a middle course is possible. Dame Mary Warnock was the chair of the committee whose recommendations led to the HFE Act (1990) and to the establishment of the HFE Authority. At face value, the Warnock Report, published in 1984, supports a very utilitarian view of the early embryo:

> *According to the majority view, the question was not, as is often suggested, whether the embryo was alive and human or whether, if implanted, it might eventually become a full human being. We concluded that all these things were true. We nevertheless argued that, in practical terms, a collection of 4 or 16 cells was so different from a full human baby or a fully formed foetus, that it might quite legitimately be treated differently. Specifically, we argued that, unlike a full human being, it might legitimately be used as a means to an end that was good for other humans.*[9]

However, the same report urged ethical respect for the human embryo and suggested that it 'ought to have a special status' under English law. This would mean that early embryos were not be used for trivial research.

Of course, use of human embryos to create stem cell cultures was not envisaged when the Warnock Report was written, nor when the HFE Act was established. This topic was however very much part of the amendment to the Act, debated in 2000 and 2001 and passed in 2001. Mary Warnock voted against the amendment in the House of Lords on the grounds that, despite her views as set out in the quotation above, she believed that to create embryos as sources of stem cells would be a step too far in their commodification. She has since changed her view and is now supportive of human embryonic stem cell research. Nevertheless, there are those who hold the position that she expressed in the Parliamentary debate. That is, they do not believe that early human embryos are human persons, but nevertheless their use to generate stem cells is so far from their natural course of development (even given that 80 per cent fail to undergo this) that it does indeed represent a step too far in their commodification. However, this is certainly a minority view in the UK, with most of the debate occurring between the proponents of the two main views set out above.

This section would not be complete without a brief discussion of the situation in the USA. During the first term of office of President George W. Bush, use of human embryos for stem cell research was banned, except that a limited number of embryos, held frozen for possible future research, were exempted. Scientists wishing to embark on this type of research therefore have some, albeit very limited, resources at their disposal. As with the ban on reproductive cloning, it applies only to federally funded laboratories. In theory, non-federally funded scientists could be very active in this area. In

---

[9] Warnock Report, 1984.

general though, the debate has been along the two main lines that were set out above, the main difference being that a larger proportion of the USA population than the UK population ascribe human personhood to the very early embryo. Nevertheless, embryonic stem cell research has some high profile supporters in the USA, including the actor Michael J Fox, Ronald Reagan Jr, son of a former Republican president, and perhaps most famously the former actor Christopher Reeve ('Superman'). Reeve, who was extensively paralysed after a riding accident (and who eventually died in 2004 from complications arising from his paralysis) was convinced that such research would one day enable him to walk again. He was angry about the federal ban, stating for example that 'bigots are delaying my recovery'. The stem cell debate even featured in the 2004 presidential election campaign, with the Democrat challenger, Senator John Kerry, publicly supporting such research (and it is interesting that a Roman Catholic should take this view), while the Republican incumbent, President George W. Bush, well known as a Protestant Christian, maintained his opposition to it.

Therapeutic cloning

Research on embryonic stem cells has the immediate scientific aims of discovering what factors maintain cells in this juvenile state where they can give rise to all other cell types, and of discovering the factors that control this formation of specialized cells. It also has the medical aim of using this scientific knowledge in providing tissues for transplant into patients in order to effect repairs. One of the problems with transplants is that of rejection, and for this reason the clinicians and scientists in the transplant team will seek tissue that is as close an immunological match to the recipient as possible. It can be readily understood therefore that using nuclear replacement cloning methods to create embryos as sources of stem cells may have significant advantages. A person who needed a particular transplant could be 'cloned' by nuclear replacement but without the aim of implanting the embryo into a womb. Instead, it would be used to generate stem cells that would be immunologically matched to the patient and thus any transplanted tissue or organ derived from those cells is unlikely to be rejected.

The difficulties involved in cloning have already been noted as has the fact that those difficulties begin at the very start of the process. Further, there seem to be particular problems with primates. Nevertheless, there have been two reports (at the time of writing), in a reliable, peer-reviewed journal, of the successful establishment of human embryonic stem cell cultures from cloned embryos. The work was done in South Korea by a joint Korea–US[10] team and was reported in February 2004 and in May 2005. In the earlier work,

[10] *Evidence of a pluripotent human embryonic stem cell line derived from a cloned blastocyst,* Hwang, W.S., Ryu, Y.J., Park, J.H., Park, E.S., Lee, E.G., Koo, J.M., Jeon, H.Y., Lee, B.C., Kang, S.K., Kim, S.J., Ahn, C., Hwang, J.H., Park, K.Y., Cibelli, J.B., Moon, S.Y., *Science*, **303**, 1669–1674 (2004); Hwang, W.S., *et al.*, *Science*, doi:10.1126/science.1112286 (2005).

the team managed to generate 30 blastocysts by nuclear transplant cloning and from those, they were successful in establishing one embryonic stem cell line. In the later paper, they report success in establishing stem cell lines from patients with degenerative illnesses. This was hailed in the media as a major breakthrough. Nevertheless, the failure rate at every stage of the process meant that large numbers of oocytes (eggs) were required in these experiments and it has been suggested that availability of oocytes will be a limiting factor if such research is continued more extensively. It goes without saying that even with the development in Korea 'therapeutic cloning' (as this procedure has been called) is still a long way in the future.

Question

Does the creation by cloning techniques of human embryos for use as sources of stem cells raise any new ethical issues that are not raised by using IVF embryos as sources of stem cells?

According to the HFEA, the answer to the question in the box is 'No'. Based on this, the Authority has recently granted two licences (to a laboratory in Newcastle-Upon-Tyne and to the Roslin Institute where Dolly was cloned) to create human embryos, and thence embryonic stem cells, by nuclear transfer. Other applications for licences are likely to follow. Most of the reactions have been very much along the same lines as the two main positions relating to embryonic stem cell research in general, with those who hold that the early human embryo is a full human person being firmly opposed to the granting of the licences. Indeed, at the time of writing, a legal challenge to the granting of the licences is being mounted by a group of lawyers. However, in addition to the clear positions set out previously, we encounter another, less definable reaction, namely that if we accept cloning technology in order to generate stem cells, then it will make easier to accept reproductive cloning. This of course is an example of the slippery slope argument, an argument that is rejected by many ethicists, including Mary Warnock, but accepted by others, including, in the USA, Leon Kass, the chair of the President's advisory committee on stem cell research.

Adult stem cells

Although embryonic stem cells have the widest potential for development into many different types of cell (because that is their function in normal embryonic development), they are not the only type of stem cell. Fully formed mammals also have stem cells, different populations of which are responsible for replenishing cells that have short lives (such as blood cells), wound healing and tissue repair. For example, the stem cells in bone marrow give rise to all the different types of cell found in the blood. They differ from embryonic stem cells in that they can only give rise to a limited range of cell types. Further, some tissues have only a limited capacity for repair. Nevertheless, there is active research on adult stem cells and especially on the process of trans-differentiation, that is the 'persuasion' of one type of stem cell to undergo a developmental pathway that is not its normal one. Indeed, there have been some notable successes with this, and it is possible that some time in the future there may be the potential to generate, from a small range of adult stem cell types, a bank of stem cells for use in tissue repair and transplant therapies.

Question

Does the use of adult stem cells have any 'ethical advantages' over the use of embryonic stem cells?

Of course, one of the factors that will influence answers to that question is one's view of the early human embryo. Those who hold an ethically conservative view of the early embryo will press for research on adult rather than embryonic stem cells, and such views have certainly been expressed publicly in the UK and USA. Further, they may argue that if more resources were put into such research, adult stem cells might turn out to have at least the same potential as embryonic stem cells and will point out some the successes reported so far. People holding opposing views will suggest that adult stem cells may have some potential but that there can be no denying the natural broad potential of embryonic stem cells. If resources are limited, it is the latter that should be prioritized for funding. This is thus just one more element in a debate that seems set to run and run.

Summary: Stem cells

- Stem cells are cells that can develop into many different cell types

- For example, bone marrow cells give rise to all the different types of blood cell

- Stem cells that occur in the early embryo have the potential to develop into all the different types of cell found in the body. Because of this they are described as pluripotent

- Stem cells that occur in a developed body, whether infant or adult, have a more limited potential – they give rise to a more limited range of cells such as blood cells. They are therefore described as multipotent

- Research is in progress aimed at using stem cells in repair of damaged tissues and organs, i.e. to grow 'spare parts.'

- Because embryonic stem cells are pluripotent they are regarded as the best source of stem cells

- Embryos for stem cells may be created by cloning techniques or by *in vitro* fertilization (the latter process creates spare embryos)

- This raises questions about the ethical status of the early embryo and about commodification of embryos

- There is also active research into the possibility of widening the developmental potential of adult stem cells

10 The new reproductive technologies

The condition of infertility is usually traumatic for the sufferers. Nobody is ill with it or usually dies from it. It's an invisible but devastating disability. For couples it challenges their role as procreators, their masculinity and femininity; it results in feelings of incompleteness, guilt and deep unhappiness.

From *Child Chat Newsletter* (Summer 1984), quoted in *A Child – At Any Cost?*, Mary Mealyea (1987)

10.1 Introduction

That the ethical debate about reproductive medicine lags way behind what is technically possible is thrown into no sharper relief than by the case of Natalie Evans and her partner Howard Johnson. Natalie was 30 when she had to have her ovaries removed as part of a successful treatment for cancer. Before this was carried out, she and Howard had fertility treatment, producing embryos that were frozen so that they could attempt to have a family at a later stage. Unfortunately, the couple split up. Natalie's only hope of having a child that was genetically hers was to have the frozen embryos transferred to her womb to try to become pregnant. Howard however refused to give his consent to this procedure, as was his right under the UK Human Fertilisation and Embryology Act (1990). Natalie embarked, with no success to date, upon a series of appeals against the judgment that she should not be allowed to use the embryos in an attempt to have a child.

The ethical issues here arise from the fact that the two parents have an equal say in the fate of embryos created by assisted reproductive technologies. This chapter will analyse and discuss aspects of the complex arguments behind this apparently unproblematic statement. Because successful assisted reproduction procedures result in the creation of an embryo, some of the ethical problems basically arise from concern about the moral status of that embryo. This aspect of the topic is discussed in detail in Chapter 11.

Introduction to Bioethics, by John Bryant, Linda Baggott la Velle and John Searle
Copyright © 2005 by John Wiley & Sons, Ltd.

10.2 Gametes outside the body

The UK was the first country in the world to pass legislation to regulate fertility treatment. The Human Fertilisation and Embryology Authority (HFEA) was set up to oversee the 1990 HFE Act by a system of inspection and licensing of clinics that offered any treatment involving the handling of gametes (eggs and sperm) outside the body. This includes donor insemination (DI), *in vitro* fertilization (IVF) and more recently introduced techniques such as intracytoplasmic sperm injection (ICSI). There are many variations and refinements of these techniques, but all are covered by the Act. There have been many challenges to its stipulations, some of which are exemplified in this chapter, and the public debate has intensified, often criticizing the HFE Act and the work of the HFEA. It has been recently argued, by the leading fertility specialist Lord Robert Winston, that the HFEA should be abolished, claiming incompetence over its decisions in those cases where new ethical dilemmas are faced, such as recent debates about sex selection of embryos and creation of tissue-matched embryos for sick siblings. Lord Winston said that the HFEA should be replaced by

> *something a great deal less bureaucratic, which doesn't inhibit research, which has a better consultation process with the public and which has a much more adequate inspection process.*[1]

In response to this, Suzi Leather, Chair of the HFEA, defended the authority, saying

> *having a regulator has given the public confidence in the infertility sector and the system of regulation.*

She also believed that it is time for Parliament to revisit the 1990 Act, because technologies have advanced so much since it came onto the statute book that parts of it are out of date. It is this increase in medical and scientific knowledge, understanding and skill, coupled with the powerful needs of involuntarily childless couples, that brings new and seemingly irreconcilable dilemmas regularly into conflict with the aging legislation, and thus into the news headlines. Concerned by the speed of medical advances, some religious leaders have recently called for the instigation of a national body to debate ethical issues.[2] Hard on the heels of criticism of the HFEA, the Christian leaders in England together with the Chief Rabbi have expressed fears that controversial rulings on matters of reproductive medicine, such as the recent granting of permission to clone human embryos for research into, for example, Parkinson's and Alzheimer's diseases, are fuelling public disquiet and are being introduced without due discussion of the underlying moral principles.

[1] BBC News, 10 December 2004.
[2] *Daily Telegraph*, 20 December 2004.

10.3 Techniques of assisted reproductive medicine

Introduction

It may be helpful, before describing what is physically involved in the basic techniques now routinely employed in the treatment of infertility, briefly to rehearse the fundamental religious objection to it, at least as defined by the Roman Catholic Church (many other denominations of the Christian faith are not opposed to assisted reproduction). Essentially the basic objection is the same as that to contraception, and concerns the separation of sexual intercourse and procreation. Any form of physical barrier between sperm and egg, such as a condom, or behaviour, such as *coitus interruptus* (withdrawal, or 'spilling the seed'), that prevents possible fertilization is deemed to be morally wrong, as it is contrary to natural law. For many Roman Catholics their church's teaching poses a great moral dilemma, because tradition also requires them not to act against their consciences. Therefore, if they believe that allowing repeated pregnancies and very large families is not what God intends for them, what are they to do? Other Christians believe that couples have a right and even a moral duty to limit the size of their families, and so the use of contraceptive methods is validated. Obviously, in order to produce a sperm sample for an assisted reproduction procedure involving gametes outside the body, masturbation is the most convenient method. The teaching of the Roman Catholic Church is that the male genital organs are for procreation and urination; therefore, any other use of them (e.g. masturbation, which is a version of 'spilling the seed') is unnatural and thus unethical. Roman Catholics are not the only people with this view. In some African cultures masturbation is believed to compromise potency, and men refrain from it. This has implications for donor insemination (DI), because men of this ethnic origin rarely volunteer to be sperm donors.

Donor insemination

Treatment of a patient by DI is the culmination of much commitment by many people. The HFE Act states that in deciding to offer treatment the clinical team should take into account the potential child's right to have a father. DI is an appropriate treatment for couples in which the man has been shown by medical and scientific testing to be sub-fertile or infertile, perhaps as a result of chemotherapy for testicular cancer, or a naturally low sperm count. It is also an appropriate treatment for lesbian couples or single women wishing to have a family. However, in the latter cases, the 'right to have a father' clearly becomes an issue: the right of the child to a male parent comes into conflict with the right of the women to become parents. Over the last

40 or so years, there has been a tilt in Western attitudes concerning what is meant by marriage and 'the family'. Although the vast majority of children are born to heterosexual couples in formally married unions, there is a growing acceptance of other forms of procreation and family life.

Ethical dilemma

Jane and Mary have been together for eight years. Their families both accept their relationship. They have their own three-bedroom house and large garden. Jane has a secure job with a high salary and Mary works part-time. They would like Mary to have a child, but do not want to involve a man. They go, through their GP, to a fertility clinic to seek DI treatment.

Question

What are the arguments for and against DI treatment of this couple?

There are three main steps in DI: recruitment and screening of donors; testing, freezing and preparing the sperm and finally insemination. Each stage is fraught with ethical problems. It is quite difficult to recruit donors, because they must undergo rigorous screening for infections such as HIV and hepatitis, and adverse familial genetic traits such as Huntington's disease and cystic fibrosis. Here, medical ethical issues arise for the donors themselves. They must then commit themselves to months of weekly donation, which involves producing samples in the clinic. They are only paid expenses for this service, so there is no financial inducement. At present, sperm donors cannot be identified, but many of their medical details have to be recorded, as it is the right of a person conceived by DI to have knowledge of certain aspects of this information about the donor. This situation is set to change, as it is thought that the right of a child born as a result of DI is stronger than the right of the donor to anonymity. There is increasing pressure for much more genetic information to be made available, and this inevitably results in fewer men being prepared to become sperm donors. Sometimes, the sperm samples do not survive the various necessary laboratory procedures required to store them in the liquid nitrogen, which preserves them almost indefinitely at very low temperatures. As a result of all these difficulties, the number of samples available for treatment is often relatively low.

Having decided to accept a patient for treatment, the clinic chooses a sample based on information kept about its donor, such as skin and eye colour, build and other physical features. Where possible, the match is made to the woman's partner. Another pitfall is that the sample may not survive the thawing or insemination procedures. However, as a fertility treatment resulting in the live birth of a baby, DI has an overall success rate of 11 per cent per attempted treatment cycle. Clinics will often warn patients that they should look upon their DI as a course of at least six cycles, rather than a one-off treatment. Very little DI treatment is available through the National Health Service in the UK and, depending upon the amount of other treatment necessary, patients can face considerable expense. All this information must be given to patients attending a fertility clinic, so that their consent to treatment is informed.

In vitro fertilization and variations

As a treatment for infertility, DI has, in many cases, been superseded by newer technologies, all of which have been made possible by *in vitro* fertilization (IVF). Previously infertile men can now be helped to become biological fathers. However, IVF is a far more invasive treatment, not only of the patients' bodies, but also of their lives, involving as it does

(a)   stimulation of ovulation during the treatment cycle (sometimes called 'super-ovulation');

(b)   semen collection, analysis and preparation;

(c)   egg collection;

(d)   insemination *in vitro*;

(e)   fertilization and embryo culture;

(f)   embryo transfer back to the woman's uterus.

At any of these stages, the treatment can, and frequently does, fail. The latest figures available give a success rate for IVF in women of all ages of about 22 per cent per treatment cycle. This falls as the age of the woman rises. There are some variations on the basic procedure, such as treating the egg in various ways to assist the passage of the fertilizing sperm, or transferring a collected egg and some specially prepared sperm back into the woman's fallopian tube. As with DI, patients are advised to view IVF as a course of treatment, but because many of the natural barriers to fertilization are removed it is statistically the more successful approach. As well as potential failures, a range of ethical problems marks each stage of IVF. Many of these, particularly those of consent to treatment, are common to issues in other medical specialisms, but some are specific to reproductive medicine. Ethical problems arise in particular in the following situations:

- if the patient has no long-term partner;
- if the prospective parents are of the same gender;
- about the origin of the gametes if donors are needed – who the donors are and what was their motivation;
- about the genetic status of the child.

As in all human situations, if something can go wrong, it may well do so. When the consequences of fertility treatment produce an ethical dilemma, it frequently seems that the debate lags behind what has actually happened. A good example of this occurred in 2004, when overworked staff and poor management in a clinic were blamed for the birth of mixed race twins to a white couple following IVF treatment.[3]

Within the last five years, the technique of injecting a single sperm directly into an egg has been developed with increasing success. This is known as intra-cytoplasmic sperm injection (ICSI), and for the patients involves all the stages of IVF. The difference lies in the considerable expertise needed by the embryologists, who manipulate the individual gametes by hand under a microscope, passing a single sperm cell through a very fine glass tube into the inside of the egg, before incubating it to produce an embryo, which is then transferred to the woman as in conventional IVF. This treatment is suitable for couples in which the man has a low sperm count, or a high proportion of abnormal sperm cells.

Ethical dilemma

It has been suggested that by bypassing the natural barriers to fertilization, genetic defects of sperm production may be passed on to a baby boy.

Question

Should a couple who have had to resort to ICSI to have a child do so, knowing that if it is a boy he may also have to use ICSI when he is old enough to have children?

---

[3] *Daily Telegraph*, 23 June 2004.

Since the first IVF baby, Louise Brown, appeared in 1978, more than 68 000 children worldwide have been born through IVF. In the UK, the NHS provides about 25 per cent of IVF treatments. If patients opt for private treatment, a single cycle of IVF can currently cost £2000–£4000, with drugs amounting to an additional £1000. The funding of IVF treatment raises a host of ethical questions. Essentially, infertility is a condition of otherwise healthy people, so some might argue that limited NHS funds are better deployed in providing, for example, more kidney machines, heart transplants or neonatal nursing care for sick babies already born. Those in favour of IVF would argue that involuntary childlessness is a condition that causes huge misery and distress and deserves treatment at least as much as for example people who have made themselves ill by smoking or excessive drinking.

Ethical dilemma

Two couples are being treated with IVF at a clinic. The early stages go well for both couples, and healthy embryos result from the culture stage. After counselling, they both decide to have two embryos transferred. However, unbeknown to anyone there has been a mix-up of labelling the embryos in the incubator, and as a result the women receive each other's embryos. For couple A there is an unhappy outcome: the woman fails to become pregnant. Couple B go on to have healthy twins. Some months later medical evidence comes to light to suggest that the babies may genetically belong to Couple A. The mistake in the IVF clinic is then realized.

Question

Which family should the twins belong to?

In 1998 IVF technologies made possible the birth of two children to Pauline Lion at the age of 55. She placed notices in local newsagents, supermarkets and newspapers to find an egg donor, spending £11 500 on the subsequent IVF treatment. More recently, an Italian woman aged 60 became pregnant with twins through donated gametes and IVF. She gave birth prematurely to a daughter, but the other twin died. Cases of this type raise the question that although men can become fathers into old age, women who have beaten the menopause by assisted reproductive techniques come in for direct or implied criticism.

Returning to the issue of the right of a child to a father, the case of Diane Blood raised a considerable ethical debate and protracted legal wrangle. Diane's husband, Stephen, fell into a coma as a result of meningitis. Told that he was not likely to recover, Diane requested that sperm be collected from him. In spite of the fact that he was unable to give consent, this procedure was performed, and the sperm frozen in storage. Stephen died without recovering consciousness. Arguing that they had agreed to start a family and that it would be Stephen's wish that Diane should use his sperm to have a child, her hopes were dashed when the HFEA ruled that this was not permissible, because Stephen had not consented to using his sperm for fertility treatment. Diane overcame another hurdle to take the frozen sperm to Belgium, where she was allowed to have ICSI treatment. Her son Liam was born in 2000, and again using Stephen's frozen sperm Diane succeeded in having a second son, Joel, in 2002. Her final legal victory came in 2003, when she was allowed to register Stephen posthumously as their father.

10.4 Designer babies

Molly Nash was born in 1994 with Fanconi anaemia, a rare genetic condition in which the body cannot make healthy bone marrow. Sufferers rarely reach adulthood. Her parents went to a treatment centre where embryos were produced by IVF and then genetically tested to ensure the absence of Fanconi anaemia and immunologically tested to ensure a tissue match with Molly. The one embryo (of the 14 created by IVF) that met both criteria was transferred to Mrs Nash's uterus in an attempt to create a possible donor sibling for Molly. The Nashes' protracted, but eventually successful, treatment resulted in baby Adam being born in 2000. Blood from his umbilical cord was collected at the time of his birth and stem cells from it have been successfully used as a bone marrow graft for Molly. This appears to be a story with a very happy ending, untrammelled with ethical problems, but critics have argued that baby Adam could not give his consent to the harvesting of his tissue, that creating him for this reason amounted to 'commodification' of human life and further that the selection procedures involved actually meant that he was a 'designer baby'.

Until 2004, the Human Fertilisation and Embryology Authority (HFEA) has allowed embryo testing only for serious genetic disorders. In several recent cases, British parents, encouraged by the Nash success, have been permitted to test embryos if the testing benefits the actual embryo, but not where the only purpose is to help another person, with no benefit to the potential child represented by that embryo. One such case is that of the Whitakers, Michelle and Jayson, whose son Charlie had a life- threatening but non-inherited blood condition, Diamond–Blackfan anaemia, named after the doctors who first described it. They applied to be allowed to test embryos so as to provide a sibling who could be a donor for Charlie. The HFEA refused on the grounds that this was not a genetic condition, and therefore the embryo itself would not benefit from the testing process. The Whitakers went to the USA, and after considerable physical and financial stress produced Jamie, who is a match for Charlie and has since provided umbilical cord blood in a bid to cure him.

This case is in contrast to that of the Hashmis. A landmark Court of Appeal ruling in 2003 allowed Raj and Shahana Hashmi to select embryos to have a baby to help their son Zain, whose inherited blood condition, β-thalassaemia, could be relieved by a bone marrow transplant. The key difference between the Whittaker and Hashmi cases lie in the fact that Zain Hashmi's condition has been linked to identified genes. Judges said the ruling should not start an avalanche of genetic testing of embryos, and that in future each case must be judged on its own merits. In the Hashmi case, testing was seen to be good for the embryo as well as for the sibling, because it would also prevent the potential child from having β-thalessaemia. However, the HFEA has since changed its stance and ruled in favour of another case like that of the Whitakers (July 2004), where no other donor could found for the sick child.

Medical advances, and the increasing number of apparent discrepancies between UK procedures and those allowed abroad, are influencing public opinion. This has led, amongst other things, to criticisms of the HFEA. Nevertheless, clinicians and scientists in many other countries look with envy upon the regulation of reproductive medicine in the UK. This does not mean that UK regulations are perfect, and one of the challenges of the 21st century will be the way in which newly available technologies are brought into the regulatory framework.

As genetic understanding increases, it is likely that more complex disease inheritance patterns will emerge.

Question

Where it can be clearly shown that a wanted child can be a stem cell donor for a sick sibling, should embryo testing always be allowed, or will producing a child in this way be treating it as a commodity?

10.5 Men and women – do we need both?

This might seem a very strange question because societies in all countries throughout the history of humankind have developed because of the relationships between men and women, but reproductive technologies in theory make it possible to do without either. Human sperm are the smallest and most highly specialized cells in the body, and over the course of evolution have developed as they have to be perfectly designed to fertilize an egg. Indeed, the array of structural and functional designs and specializations seen in the sperm cells of mammalian species is quite amazing: the musk shrew sperm have an enormous acrosome (a structure containing powerful enzymes to help the passage of the sperm head into the cytoplasm of the egg), which dwarfs the rest of the cell; hamster sperm have hooked heads and the American opossum sperm form pairs in which they swim through the female reproductive tract with their tails in synchronized motion! However, the only actual part of the sperm that is necessary to produce an embryo in any species, humans included, is the nucleus, containing the genetic information, so if one artificially introduces that part into an egg, as for example in the process of ICSI, all the highly evolved specialist structures of the sperm – its acrosome, the mid-piece, packed with mitochondria for swimming power, and the flagellum – are all rendered unnecessary, as are all the parts of the man's reproductive tract that are specialized for the differentiation of the mature sperm cell out of the stem cells from which they arise.

Eggs, in contrast, are the largest cells in the body, and they too have important membranes and vestments around them, which play a vital role in the passage of the egg from the ovary down the Fallopian tube, and in the process of fertilization. Eggs also contain special cytoplasm or 'yolk', which plays a vital role in the first few cell divisions of embryonic life. Just like sperm, these specialisms have evolved over time to ensure reproductive success, but the vital part is the genetic component of the egg. As ICSI and cloning technologies have shown, it is entirely possible to introduce genetic material into an egg and create an embryo, which can develop into offspring. Animal experiments have shown that the genetic component of an egg can successfully be placed into another, enucleated egg, and that offspring can result from this. These are all female. This suggests that sperm are not needed at all, and since it is the sperm that determines the gender of mammalian offspring, these experiments suggest that reproduction is technically possible without males. It is reassuring to know that although it is quite successful in frogs this means of reproduction has a low success rate in mammals, and is never likely to be allowed in humans. Although it is not at present possible to create an embryo without the membranes and cytoplasm of the egg, the rate of advance of the science is such that even this biological hurdle may in time be overcome, opening up the prospect of at least creating an embryo from basic genetic material, and in time, with improved incubation methods, a full-term offspring.

This distinctly unpleasant spectre is what could face us as we continue to manipulate and control natural reproductive systems. As the Archbishop of Westminster, Cardinal Cormac Murphy-O'Connor, England's leading Roman Catholic, has said,

> *the issues which the new technologies have thrown up touch on the very source and mystery of life. We need an ethical rigour capable of meeting the challenge.*[4]

10.6 Conclusion

The ethical issues surrounding human reproduction are at the same time fundamental and extremely complex: this makes them difficult, if not impossible, to be completely right or wrong about. The controversies give rise to real dilemmas because of technical difficulties, uncertain outcomes and variance of principles. Critical reading and frequent discussion help to inform a person's views, and this is very important because laws that exist on such matters as assisted reproduction and embryo research should not be enacted without wide public consultation and debate. However, the views of society as a whole may not always reflect the most ethical actions; for example, there is still a majority in favour of capital punishment in the UK but we do not do it. On the other hand, a majority opposed the 2003 war in Iraq but the UK still engaged in it, a decision that many believed to be highly unethical. However, changes in the law are sometimes made to reflect shifts in people's opinions about what is acceptable. The relationship between social morality and the legislature is not static; indeed, one informs the other, as can be seen in the current argument between researchers in reproductive medicine and the HFEA. Engaging in debate and voicing opinion is particularly important when new guidelines are being drawn up. The HFEA regularly puts out public consultation documents on such issues as sex selection and therapeutic cloning, and the questions raised by these matters are discussed by staff and patients at treatment centres, religious groups, political gatherings, academic departments and many others. Above all, it is most important for everyone to keep trying to come to terms with these difficult moral problems, by thinking carefully about their significance not just for themselves, but also with empathy for those more directly affected.

---

[4] *Daily Telegraph*, 20 December 2004.

11 Embryos, foetuses and abortion: issues of life before birth

Leigh's coup is to transfer the sympathy from the pregnant woman to the abortionist herself – a brilliant upending of the traditional stereotypes and pieties ... the film plainly shows the squalid hypocrisy of Britain before the Abortion Act.

From a review in *The Guardian* (7 January 2005) by Peter Bradshaw of Mike Leigh's film *Vera Drake* (2004)

We forthwith acknowledge our awareness of the sensitive and emotional nature of the abortion controversy, of the vigorous opposing views, even among physicians, and of the deep and seemingly absolute convictions that the subject inspires ... Our task, of course, is to resolve the issue by constitutional measurement, free of emotion and of predilection.

Mr Justice Blackmun, giving the opinion of the Supreme Court of the United States in *Roe vs Wade*, 1973

11.1 Introduction

When we look at a newborn baby we have no doubt whatsoever that we are looking at a human being. We have a sense of wonder and awe. We exclaim 'Look at the tiny hands – they are perfect'. Almost immediately this certainty that the baby is a person is reinforced as we speculate which of the parents it most looks like. There is an instinctive desire to protect the child, to shield it from harm, to provide for it and nurture it. Of course we recognize that there are many years ahead of growth and development before this baby becomes a responsible human adult. However, whether a person is one day or 90 years old, the law gives them the same protection and confers upon them the same value and dignity. Human life recognizably begins at birth and

Introduction to Bioethics, by John Bryant, Linda Baggott la Velle and John Searle
Copyright © 2005 by John Wiley & Sons, Ltd.

ends at death. Until the second half of the 20th century a baby born prematurely had little chance of survival and the moment of death was easy to determine. We shall look at the dilemmas modern medicine has created around death in Chapter 12. However, technological advances have given rise to no fewer problems at the beginning of life. Babies born as early as 24 weeks of pregnancy can survive because of modern neonatal intensive care, often with a good quality of subsequent life. However, what of a baby's status before it is born? Should the unborn child or foetus have the same rights and protection in law as a baby who is born at term or who is, after a very premature birth, optimistically being looked after in an intensive care unit? If so, when should those rights be conferred on it – from the moment of conception or fertilization, or when a woman realizes she is pregnant, or when she feels the baby move, or when? In present day UK law, the foetus has no *rights* until it is born. Most societies put the rights of a pregnant woman above those of a foetus, the argument being that the mother is a person and has responsibilities whereas, before birth, the foetus has no legal standing as a person.

Within the usual understanding of the concept of 'rights' comes the acceptance that with rights come responsibilities. Clearly this cannot be the case for an unborn child (or for that matter for a baby), but in a biological sense the struggle to maintain life, to survive, might be considered to be the responsibility of the child as a member of the human species. People, as far as we know, are the only extant species able to rationalize, and we have a basic instinctive drive to protect life. This has been termed the 'presumption in favour of life', and its consequences are often keenly argued in medical ethics generally, but sharply focussed at the beginning and end of life. However, the status of the unborn is morally problematic, because decisions often hinge upon agreement about when human life begins.

These questions and the debate they engender have greatly intensified during recent years in relation to the issues of embryo research and abortion – the termination of pregnancy. This is an important debate. It concerns issues about which there are strong feelings. It is thus against this background that this chapter consider issues at the beginning of life, including the questions about when human life begins, about the moral status of embryos and of foetuses and about abortion.

Question

When does an unborn child become a human being? Should it be given the same protection in law as a newborn child?

11.2 The early human embryo

It is very easy to ask the question 'when does human life begin?'. For some people the answer is also easy, i.e. when the sperm fertilizes the egg. For others it is more difficult. Throughout history passionate debate has raged on questions such as at what stage the duty to protect human life begins, what is entailed by or involved in our duty to protect human life, when it is justifiable to interrupt the fulfilment of human development and so on. At one time the Roman Catholic Church believed that the soul entered the body 40 days after conception for a male foetus and 70 days for a female! The teaching authority of the Roman Catholic Church and the views of many conservative protestant Christians currently hold that life begins at fertilization, and that a human life is distinguished from that of animals because humans are spiritual beings (a concept that we discussed in Chapter 4), a quality that distinguishes us from the rest of the animal kingdom. But when does this spiritual element actually arise? Alternatively, to put it in terms more generally understood by those who have no religious faith, at what stage does the embryo or foetus become a person? Can we ascribe human personhood to the eggs and sperm, or even to their progenitor cells in the ovaries and testes? Or does personhood start at any of the following stages:

- when the sperm and egg first make contact?

- as the sperm penetrates the outer layers of the egg?

- when the sperm nucleus lies alongside that of the egg?

- when the genetic components of the sperm and egg finally unite?

- during the first cell divisions of embryonic life (see Figure 11.1 below)?

- at the blastocyst stage?

- at the 'primitive streak' stage (when the beginnings of the nervous system are first laid down, and when it ceases to be an embryo and becomes a foetus)?

- when the mother first feels the foetus move inside her?

- when the foetus is capable of independent life?

The question of the status of embryos was forced into the public domain in 1978 when Louise Brown, the world's first test tube baby, was born. Her mother's egg was fertilized with a sperm from her father in the laboratory where it was allowed to grow for a few days. It was then implanted back into her mother's uterus and the pregnancy proceeded to term when Louise was born. However, this embryo was not the result of a unique feat of embryology. The technique required several eggs to be fertilized in the laboratory and hence the production of several embryos. What was the status of these 'spare' embryos? Were they human beings? What should be done with them? What protection should the law give them? At the time there were two views in answer to these questions and the debate continues today. These views are

- human life begins at fertilization;

- we cannot be sure when human life begins but experimentation and other work on the human embryo may reasonably be conducted up to 14 days after fertilization (as in the terms of the Human Fertilisation and Embryology Act, 1990).

Status of the embryo: human life begins at fertilization

This view is held by so-called pro-lifers, many of whom are Roman Catholic and conservative protestant Christians. There are two main planks to their argument: first that fertilization is a specific occurrence (actually it is a series of events which takes several hours), and that it is definite – as definite as birth or death; second, the fertilized human egg contains the full genetic complement of the new human being: that is, everything that is genetically necessary for its future growth and development not only as a baby but also into an adult. The pro-life view is summarized in the Encyclical Letter *Evangelium Vitae*, of Pope John Paul II, issued in 1995: *'procured abortion is the deliberate and direct killing by whatever means it is carried out of a human being in the initial phase of his or her existence, extending from conception to birth'*. Such a view means the following:

- No research can be ethically carried out on the early human embryo, including into the causes of genetic disease. To do so would be to use a human being instrumentally, as a means to an end.

- Abortion is always wrong and pregnancy must be allowed to continue to birth.

- Infertile couples can never have a child of their own (this issue was discussed in Chapter 10).

Question

For you, is it ethical or unethical to fertilize human eggs outside the uterus for the treatment of infertility?

Status of the embryo: the 14 day approach

The egg and the sperm each contain half of their generator's DNA. When the sperm and the egg meet at fertilization each fertilized egg (zygote) then contains a full complement of human DNA. For the first few days the zygote floats freely first in the Fallopian tube and then in the uterus. The genetic identity of a new individual is established at *syngamy*, about 30 hours after the initial encounter of the fertilizing sperm cell with the egg membrane, and it is at this stage that *cleavage*, or embryonic cell division, begins. Figure 11.1 shows the early development of the human embryo.

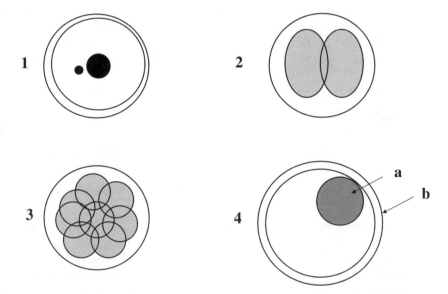

Figure 11.1 Diagrams of the early development of the human embryo. Reproduced by permission from Bryant, J. and Searle, J. (2004) *Life in Our Hands*. Inter-Varsity, Leicester.

1   Shortly after the penetration of the egg by a single sperm cell, the two nuclei, each containing one set of genes, are clearly visible. Note that although the two nuclei contain equivalent amounts of DNA, the egg cell nucleus is much bigger than the sperm cell nucleus. The complete fusion of the two nuclei and the activation of the fertilized egg to divide will take about ten more hours.

2   The first cell division has taken place (note that the nuclei are not shown in diagrams 2–4).

3   After two more rounds of cell division the embryo is at the uncompacted eight-cell stage. The embryo is still 'indeterminate' and it is thus possible to remove one cell for genetic testing (as described in Chapter 7).

4   The blastocyst stage. The embryo now consists of an inner mass of cells (a), which will, if the embryo implants, become the embryo proper, and an outer layer of cells (b), from which the placenta and foetal membranes will be derived if a pregnancy is established. Stem cell cultures may be established from the inner cell mass (blastocoel).

Thus, the embryo undoubtedly has its own unique version of the human genome, giving it the potential for a life within humankind. But is the moral status of the embryo based just on genetic uniqueness? What then is the status of certain aberrantly formed embryos? For example, it has been known for some years that human eggs can undergo cleavage in the absence of fertilization. This condition is known as parthenogenesis, and in an experimental situation development beyond the primitive streak stage has been observed from unfertilized mouse eggs. Can we ascribe personhood to a parthenogenetically dividing embryo? Further, if the possession of a unique human genome is a major criterion, what can be said about cloned embryos?

At the blastocyst stage there are a number of possible outcomes. It can divide into two and become identical twins: this is of course the natural production of clones. It can develop into a single foetus with its supporting system of the surrounding membranes and placenta. However, about 80 per cent of fertilized eggs simply do not survive. There is a huge natural wastage. They are fertilized eggs but in the natural order do not fare as foetuses.

Question

Can these failed embryos be described as human beings?

At around the stage of 30–60 cells, which is at about 10 days, the blastocyst begins to attach itself to the wall of the uterus. As this happens certain hormonal changes occur in the mother, she misses her monthly period and begins to 'feel pregnant'. At 14 days the cells become more organized and a strip of cells called the primitive streak forms, from which the nervous system will develop. By 10–11 weeks organs are formed and early limbs are also present.

In summary, between fertilization and 14 days after fertilization the survival of the fertilized egg is in doubt and many do not survive. Survival depends on the embryo imbedding itself in the wall of the womb. Furthermore, the early cells may develop either into a baby or the placenta, which is discarded at birth. Finally, there is no evidence of a nervous system before 14 days and a functioning nervous system is an essential part of our humanity.

11.3 Embryo research

It was obvious after the birth of Louise Brown that test tube babies or assisted reproduction would become a standard option in the treatment of infertility. The ability to fertilize eggs in the laboratory and grow embryos also opened up many other possibilities: research into the causes of infertility, miscarriage and genetic disease, identifying genetic disease and selecting only healthy embryos for re-implantation into the uterus. During the 1990s other developments occurred, such as the use of embryos to produce stem cells. The government set up a committee, chaired by the distinguished philosopher Mary Warnock, to examine the science and the ethics of these possibilities and make recommendations. Its report, known as the Warnock Report, was the basis for the legislation that currently governs this area of biomedical research and practice. The Report adopted the '14 day' approach to the human embryo. It is a classic piece of principled pragmatism. It recognized that the early human embryo was indeed human material but could not yet be regarded as a human person because first many embryos in nature never reach implantation and second the rudiments of the central nervous system are not present. To give it the same protection as a human person would be to prevent any research into diseases that cause much suffering. However, scientists were not to be free to do whatever they liked with human embryos. Their activities are governed by the Human Fertilisation and Embryology Authority (HFEA) set up by Act of Parliament in 1990. There was a minority report published around that time that took the more traditional view that human life begins at fertilization.

The HFE Act has five main provisions:

1   All activities related to human reproductive technologies are supervised by the HFEA and may only be conducted under a licence granted by the Authority.

2   Embryo donation and donor insemination are allowed under licence. The Act also sets out who are the lawful parents of children conceived by artificial reproduction and the confidentiality arrangements about the genetic parents, although this is currently under review.

3   Embryos and gametes may be frozen. They may be used in the future with the consent of the donors. They are destroyed after 10 years.

4   The legislative arrangements for surrogacy.

5   Research on human embryos may be carried out up to 14 days after fertilization under licence for the following purposes:

   •   promoting advances in understanding infertility

   •   gaining knowledge about congenital disease

   •   the development of contraceptive methods

   •   detecting gene and chromosomal abnormalities.

As new developments arise the HFEA rules on whether or not they may be pursued, granting the appropriate licence (some examples are also discussed in Chapter 10).

Question

The Warnock Committee's recommendations depended heavily on our scientific knowledge about fertilization and the development of the human embryo during the first 14 days. How far should scientific knowledge contribute to making ethical decisions?

The whole debate about whether or not embryo research is morally acceptable really centres on the definition of an embryo, and this is again related to the question of when life begins. In this respect, the arguments *against* research using embryos are similar to those against abortion. For those people who believe that a unique human person comes into being at fertilization, any procedure that interferes with the normal development of the embryo is unacceptable because they see that embryo as having the same rights and interests as any other child or adult.

Experimental procedures on human embryos are only permitted under the law for the first 14 days after the mixing of the gametes – in other words before the appearance of the primitive streak. Even then, under the law, certain types of research on human embryos are actually prohibited. These include

- replacing a human embryo in an animal

- nucleus substitution – this procedure comprises removal of the nucleus of an embryonic cell and replacing it with the nucleus taken from the cell of another person or embryo

- altering the genetic structure of any cell while it forms part of an embryo

- cloning of human embryos for the purposes of infertility treatment.

Other than in these circumstances, it is only after this period of 14 days from fertilization that protected human life legally begins. An embryo that has been experimented upon cannot be maintained *in vitro*, frozen, or replaced into a woman after that time, and it must be destroyed.

> Recent reports suggest that there is real hope of improve-
> ment in mobility for patients with spinal injuries, following
> treatments arising from stem cell research.*
>
> ## Question
>
> Should more embryos be made available for research into
> this sort of treatment? If so, where they should they come
> from?
>
> * See Chapter 9.

Recently, the HFEA has granted licences for the use of human embryos in stem cell research.[1] This was a significant moment for British science, and it has opened the door for research into a wide range of hitherto incurable diseases. This was discussed more fully in Chapter 9. It has long been held in the UK that the sensitivity of feeling about human embryo research is such that the development of any new technique must be strictly regulated. Research licenses are only granted when it can be demonstrated that the researchers will add to the body of knowledge, and in doing so gain technical competence. Only then can they be allowed to use a new technique in the treatment of patients. It is with this careful and painstaking approach, coupled with on-going public debate that big issues – and embryonic stem cell research can surely be said to be the most significant in recent years – can be approached responsibly. There is no doubt that research will continue to strive to improve on nature: indeed, discussion of the elimination of 'undesirable' genes is already one of the consequences of the Human Genome Project, but this should always be tempered with questions such as 'What actually is meant by undesirable?'.

11.4 Abortion

Abortion was unlawful in Britain until 1929, when an Act of Parliament was passed under which abortion was not a crime if it was performed in good faith to *save the life of the mother*. This was an important piece of legislation because it recognized the need to prioritize ethical principles. Under some

---

[1] *The Guardian*, August 12, 2004.

circumstances, if a pregnancy is allowed to progress to childbirth at term there is a clear risk that the mother may become seriously ill or in some cases die. The Act recognized that, while the lives of both the mother and the foetus should be protected, if there was a choice between the life of one or the other the life of the mother should take priority over that of the foetus.

However, many women continued to get pregnant but did not wish to have the baby. Their only recourse was to the so-called 'back street abortionist'. Generally such a practitioner was unqualified and the operation was carried out clandestinely, without proper controls of infection and hygiene. The mother was therefore exposed to serious risks to her own health. Often her reproductive organs were infected so that she was unable to become pregnant on another occasion. There was also a definite mortality rate from these unlawful operations.

A further development took place during the 1960s. This was the decade during which much of the restraint of former generations on sexual activity began to slacken. It was paradoxical that while this followed the introduction of effective oral contraception many women still became pregnant when they did not wish to do so, both within and without marriage. David Steel's 1967 Abortion Bill was a proper attempt to regulate abortion and protect women. The Bill was passed by Parliament. Together with the Infant Mortality Act, the 1967 Abortion Act made abortion legal up to 28 weeks from the first day of the last menstrual period, and allowed pregnancy to be terminated if two doctors, acting in good faith, agreed that

1   the continuance of the pregnancy put the woman's life at greater risk than termination,

2   the continuance of the pregnancy put the woman's physical or mental health at greater risk than termination,

3   the continuance of the pregnancy risked the physical or mental health of any existing child(ren) of the pregnant woman, or

4   there was a significant risk that if the foetus was born it would suffer from a serious physical or mental handicap.

The upper limit of pregnancy at which an abortion could be performed was 28 weeks. There were two important additional clauses in the Act that allowed an abortion to be carried out for reasons other than a risk to the life and physical health of the mother. First, was there a risk to the mental health of the mother? This of course raised the question as to what actually was a 'risk to the mental health of the mother'? Was this a history of serious postnatal depression? Did it include a mother's view that she simply could not cope with having another child for whatever reason? Not unreasonably, the doctors concerned with individual patients were left to make that judgement.

Second, abortion was allowed if there was 'a substantial risk that if the child were born it would suffer from physical or mental abnormalities as to be seriously handicapped'. This was new territory and raised serious questions not only about what constituted being 'seriously handicapped' but also about the value society placed on people with mental or physical disabilities. Was the Act saying that it would be better if such people were never born? What did that say to disabled people? On the other hand, there are abnormalities where there is little or no possibility of the child surviving for any length of time or having a life of severe difficulty and suffering. Once again, the decision was left to the judgement of the doctors.

In America, a similar turning point to the UK's Abortion Act was the court's judgement in the case of *Roe versus Wade* in 1973. As a result of this judgement, no individual state in the USA had the right to restrict the availability of abortion in the first six months of pregnancy. Individual states did however retain the right to prohibit abortion in the last three months of pregnancy except when the mother's health was jeopardized, in which case a prohibition was not legal. However, the whole topic is now very politicized in the USA and there are beginning to be differences between states in the interpretation of the judgement in *Roe versus Wade*. This is especially seen in relation to abortion in the second trimester, where there are differences of opinion as to what constitutes medical grounds for termination of pregnancy.

Therapeutic abortion remains illegal in some countries such as the Republic of Ireland, Portugal and some parts of South America. In countries where abortion is legal, the question is not so much whether it is right or wrong, but under what circumstances it is justified, as already mentioned in respect of the UK and USA. Orthodox Jewish law allows the killing of a foetus in order to save the mother until the birth of the head and there have also been some 'partial-birth' abortions in the USA. In some Eastern European countries abortion has been seen as a substitute for contraception

The result of the 1967 Abortion Act in the UK was that while it protected women from the risk to their health and in some cases their life from either a pregnancy or a back street abortion, it rapidly extended the grounds for the lawful termination of pregnancy. In the first 20 years of the Act's operation in England and Wales three million abortions were carried out. Currently in the USA it is estimated that one and a half million abortions are performed every year. The main ground for abortion now is that the woman does not wish to have the baby. How did this 'abortion on demand' come about?

The 1967 Act came at a time when the former deontological basis of ethics – not least sexual ethics – was being questioned. This was the decade in which homosexual activity between consenting adults was decriminalized. Later, sexual unfaithfulness had no longer to be proved for a spouse seeking a divorce. Divorce could take place because the marriage had broken down irretrievably. It was also the time when the modern concept of 'rights' was developing rapidly. So a woman had a right to decide whether or not she wanted to have the child she was carrying. The law gives this decision to the mother. The father has no rights in the matter at all.

The pro-life constituency has remained deeply opposed to abortion. In the USA their voice is particularly influential at present. It is said that it played a key part in the re-election of George W. Bush as President in November 2004. Their argument is twofold:

(a)   the foetus is only temporarily part of the mother who is carrying it. It is not, like other organs, part of her. Indeed, it is genetically distinct from her and will grow through prenatal life and childhood into an independent adult.

(b)   the foetus is human life – albeit not yet fully developed – and is therefore entitled to the same protection as other human beings.

This situation changed with the passage of the 1990 Abortion Act, which stipulates no upper limit for gestation in cases where there is a risk of foetal abnormality, but sets a limit of 24 weeks in cases where the woman's mental health may be affected. If a woman requests a termination of pregnancy, two doctors are needed to certify that one of the clauses of the Act applies to her case. The father does not need to consent to the termination. The doctors also have to notify the Department of Health that they have performed the procedure, and give information about the circumstances of the case. Over 100 000 legal abortions are carried out in the UK annually. It has been estimated that worldwide about 50 million abortions are performed per year, and more than 40 per cent of these by unsafe methods.

Ethical dilemma

Sue became pregnant by accident. She wanted an abortion, but her partner Jim wanted her to have the baby.

What are the arguments for and against only the woman having to give consent to a termination of pregnancy?

Some antenatal diagnostic tests that may help the parents to decide whether to terminate a pregnancy do not give reliable results until 18–20 weeks of gestation. Laws that only allow abortions before this time may not give the parents an opportunity to make an informed decision to terminate the pregnancy. However, the 24 week limit for abortion is determined, in large measure, not by the time at which the foetus becomes viable as a morally relevant cut-off point, but by the mixed strategy view, according to which the foetus has a variable moral status. Initially, the foetus is comparable in most moral respects with a piece of body tissue or an organ, and in later stages of development it is viewed in most morally relevant considerations like a newborn baby.

Some people with conservative views about abortion, the 'pro-lifers', often do not discriminate between first and third trimester foetuses, and wish to outlaw all abortion, even that carried out in the first weeks of pregnancy before all the neural structures allowing consciousness have developed. The essential belief of these people with strong anti-abortion, sometimes called pro-life, views is that the foetus is, from the moment of conception, morally on a par with an adult human being, and that killing a foetus is therefore equivalent morally to killing an innocent person.

At the opposite extreme, some people with liberal, pro-abortion, sometimes called pro-choice, views argue that women should have the unquestioning right to terminate an undesired pregnancy at any stage. The essential belief of this group is that the foetus is, from conception until birth, morally on a par with a part of the woman's body, and that the woman has complete jurisdiction over it.

Mixed strategy adherents hold a view somewhere between the pro-life and pro-choice, or conservative and liberal standpoints on abortion. They accept that people have a duty of protection towards the embryo, which increases with increasing developmental complexity, and that early abortion is less objectionable than late abortion. This lobby has put forward the view that the present law could be adjusted such that far more stringent conditions are imposed as the pregnancy proceeds. Further, if there is no diagnosis of foetal abnormality, an earlier time limit for a requested abortion than exists under the present legislation could be imposed. The most contentious standpoint is the moral justification for later therapeutic terminations of pregnancy when there is risk to the life of the mother. Many holding a mixed strategy view make this exception for abortion because they see it as comparable with killing a hapless victim from the street to save the life of the mother (for example by taking organs from the victim). The view that the death of the mother is morally weightier than any other unwanted event in her continuing life has also been called into question. It is arguable that some kinds of fate in life are at least as morally serious as death, and should therefore carry at least equal moral relevance. The philosopher John Stuart Mill (1806–1873) argued that execution was a kinder form of punishment than some alternatives. Is it right to assume that human life has a value that is independent of its actual content? This question is discussed in Chapter 12.

Similarly, therapeutic abortion on the grounds of risk of injury to the physical and mental health of other children of the pregnant woman may also be seen as morally unacceptable to the mixed strategy adherent, who would disagree with the liberal view that the foetus is morally on a par with a part of the body. It could therefore be argued that late termination on these grounds constitutes a social rather than a medical reason for abortion, and so might be catered for by social and not medical intervention. The debate thus continues.

12 Decisions at the end of life – when may I die and when am I dead?

For everything there is a season, a time to be born and a time to die; a time to kill and a time to heal

Ecclesiastes Chapter 3, verses 1–3

An extraordinary account of an extraordinary love

Melvyn Bragg, commenting on the book *Wrong Rooms*
by Mark Sanderson (2002)

12.1 Introduction – two important examples

Charlotte Wyatt was a small baby whose lungs, kidneys and brain were severely damaged. The doctors looking after her believed that she had no prospect of getting better, and therefore if she got worse life-prolonging treatment, such as artificial ventilation (a life support machine), would not benefit her. It should not therefore be used. Charlotte's parents wanted the doctors to do everything possible to save her, including using a ventilator, if she got worse. The English Courts were asked to decide between the doctors and the parents. On 7 October 2004, a judge, Mr Justice Hedley, came to a clear view, 'I do not believe any further aggressive treatment is in her (Charlotte's) best interests. I know that that may mean that she may die earlier than otherwise she might have done but in my judgement the moment of her death will only be slightly advanced'.

In this case there were two key factors. First, Charlotte was a baby. She was unable to make a decision for herself. In medical and legal terms, she 'lacked capacity'. It was argued that her parents rightfully exercised capacity on her behalf. However, if Charlotte had been over 18 years of age, nobody would have had any right to exercise capacity on her behalf. Second, the question was 'what was in Charlotte's best interests?'. In other words, the decision was about the balance we discussed in Chapter 2. Would further treatment benefit her or would it simply add to her suffering? The judge decided that further treatment could not benefit her and therefore the doctors were not required to give it, despite the wishes of the parents. He was in line with several decisions made by the courts over the preceding 13 years. We shall look at these in more detail later in the chapter.

Mark Sanderson was a journalist whose book *Wrong Rooms* was published in 2002. His Australian boyfriend, Drew Morgan, was dying from a malignant melanoma. Drew's suffering was so awful that Mark eventually smothered him with a pillow and killed him. Mark said that what he had done was not murder but rather an act of love. This story is one of many over recent years where somebody has killed a person they love. There have also been cases where doctors have done it to a patient. As in the case of baby Charlotte, there are two key points in every case. First, the terminally ill person was suffering terribly. Second, the person who killed them acted out of compassion – they wanted only to relieve the suffering. Usually the ill person asked them to do it; they consented to it. What happened was voluntary euthanasia.

Question

Is compassion an adequate reason for ending the life of a person who is suffering terribly and who has no prospect of recovery?

12.2 How did we get here?

We tend to take for granted what doctors can do in the 21st century. However, many diseases that even 20 years ago were either fatal or rendered people seriously disabled can now be treated and either cured or at least the person provided with a long period of good quality life. This has happened because of advances in surgical operations, drugs and technology. Sometimes the treatment itself has serious side-effects, which may themselves be fatal. For example, modern chemotherapy has enormously proved the outlook for some cancers, but the drugs used also make the patient very prone to serious infection. When an infection occurs the body's defence mechanisms have been temporally been put out of action by the drugs so the infection can become overwhelming and kill the patient. The doctors then have to introduce a whole load of other drugs and sophisticated technology to try to save the person's life. Furthermore, although modern treatment has greatly reduced the death rate from many diseases it has not abolished it in many of them, so when the doctors embark on the treatment they do not know whether the person is in the group that is not going to get better. When this becomes clear, should the doctors go on and on trying to cure the patient when there is no prospect of being able to do so? Anyway, how do they tell? In medical terms, when is further treatment futile?

For some patients the terminal stages of their illness can be very difficult, not only because of pain but also because of other unpleasant symptoms such as breathlessness, anxiety, incontinence, sickness and sleeplessness. It is not surprising therefore that when this occurs some people simply cannot cope any more and knowing that they are going to die in any case want their suffering brought to an end by having their lives ended. This of course is a very drastic measure, so the question has to asked whether or not these unpleasant symptoms can be controlled. There are therefore several issues that have to be addressed. These are the following.

- What is euthanasia?
- Arguments for and against voluntary euthanasia
- When should medical treatment either be withheld or withdrawn?

12.3 What is euthanasia?

The debate about euthanasia is often muddled because people use language loosely. The word 'euthanasia' comes from Greek and means a 'quiet and easy death'. In this sense most of us would opt for euthanasia in that we do not want our dying to be too difficult or to take too long. The harsh fact is that it will happen one day, but let us hope that it will be as easy as possible. Sometimes people say 'I hope I pass away in my sleep' or 'I know it would be very traumatic for my family but I would prefer just to drop dead'.

So what is 'voluntary euthanasia?' *Voluntary euthanasia* is

- the deliberate ending of a person's life
- at their request
- because that person finds their illness or disability intolerable.

It contains two important assumptions: first that someone else will do it – usually a doctor by administering a lethal dose of a drug; second that the suffering is intolerable and cannot be relieved.

Involuntary euthanasia is something different. It is

- the deliberate ending of a person's life

- without their request

- because some other person or party considers that their life is intolerable or its quality is not worth having.

Over the last 30 years there have been several attempts in Britain to make voluntary euthanasia lawful. At the time of writing (November 2004) there is another attempt to do so. Parliament is considering the Assisted Dying for the Terminally Ill Act 2004, put forward by Lord Joffe. A similar bill was debated in 1976 and the House of Lords published a report on the subject in 1994. There have also been several cases before the Courts from terminally ill patients seeking the right to have their lives ended. One example was Diane Pretty. She had motor neurone disease. This causes progressive paralysis of the body but the mind remains alert and the person has normal sensation. Diane Pretty was 43 and she requested the courts in England to allow her husband to assist her to commit suicide. This request was refused, as it was on appeal by the Law Lords in England and the European Court of Human Rights. The European Court ruled that by refusing to grant leave for her to be helped to commit suicide, the English courts were not violating her human rights. She subsequently died in a hospice.

Another case, which attracted a great deal of media attention, was that of Reginald Crew. He also had motor neurone disease. He and his wife, Win, went to Switzerland, where a person who is considered to be rational may be assisted to commit suicide under certain circumstances. A nurse gave him a lethal dose of sleeping drugs, which he took and died as a result.

Both supporters and opponents of voluntary euthanasia use cases such as these. In September 2004 a letter appeared in *The Times* newspaper from six distinguished philosophers supporting Lord Joffe's bill. A week later a letter appeared from four equally distinguished philosophers opposing it. So what are the arguments for and against making voluntary euthanasia lawful in England and Wales as it already is in the Netherlands and Belgium?

12.4 The arguments for voluntary euthanasia

There are three main arguments put forward to support making voluntary euthanasia lawful:

- Openness

- Necessity

- Autonomy.

Openness

Relatives often suspect that 'the doctor helped Grandma on her way'. The story goes something like the following. Grandma was terminally ill and had reached the last stage of her life. She was restless and uncomfortable. The doctor came, gave her an injection and she died in her sleep. The assumption is that the doctor knowingly gave Grandma an overdose of a drug to ensure that she slipped away quietly. Such an act is unlawful in Britain but the fact is that doctors recognize that while their intention in giving a drug like morphine or heroin to a dying person is primarily to make them comfortable the person does often die quietly in the sleep the drug induces because breathing is depressed. Indeed, such practice has almost certainly been seen as 'good medicine' for many years. However, although the British Medical Association is officially opposed to voluntary euthanasia, there is a lot of evidence that about half of the doctors in the UK would like the law to be changed. Indeed, anonymized surveys have shown that some doctors have actually complied with a person's request to end their lives. The argument is therefore that the law is out of step with current compassionate medical practice.

Necessity

Necessity is emotionally a powerful argument, which says that if a person with a terminal illness is suffering great pain and distress, why not help them to die? In this way their suffering will be relieved. Because they are dying, it is cruel to allow them to go on suffering. The motive is compassion, its supporters often saying such things as 'you would not let a dog suffer like this so why do we let human beings go through it?'.

Autonomy

We saw in Chapter 2 that the concept of 'rights' has become very important in 21st century Western society. Personal autonomy, the right to decide for myself what is best for me, plays a key role in both private decisions and public policy. In relation to voluntary euthanasia the argument is this: do we accept the right of human beings to decide how and when they will end their lives? 30 years ago a play was produced in the West End of London called *Whose Life is it Anyway* by the playwright Brian Clarke. The main character, Ken Harrison, has broken his neck. He is paralysed from the neck down and has no prospect of recovering. He wants to die. His doctors want to keep him alive. One line in the play neatly summarizes the autonomy argument, when Harrison says 'I have coolly and calmly thought it out and I have decided that I would rather not go on. Each must make his own decision'.[1]

Questions

If voluntary euthanasia is made lawful in England and Wales, who should administer it? Are there any problems if doctors do it?

12.5 The arguments against voluntary euthanasia

The triad of openness, necessity and compassion are powerful ones in favour of voluntary euthanasia. But what are the arguments on the other side? Two questions arise:

*   Is it necessary to kill a person in order to control their pain and suffering?

*   Is there a downside to autonomy?

Controlling pain and suffering

The growth of the hospice movement over the last 40 years has had a major impact on the care of terminally ill people. Hospice staff, whether in a ward or in the community, are experts in controlling pain and other distressing symptoms. They help dying people and those close to them to come to terms with what is happening and within the constraints of their illness make the very best of what life is left to them. They are also very good at pulling together all the various caring agencies as well as providing volunteers to keep life as normal as possible. There is much evidence to show that in 99 per cent of people dying, for example from terminal cancer, pain and other distressing symptoms can be controlled. Even in the one per cent in whom this may be difficult, hospices argue that they can still provide a reasonable quality of life over the last weeks and days of life. The argument here is why kill somebody when you can offer them the opportunity to make the very

---

[1] A new production of the play opened in London early in 2005, this time with a female leading character.

best of what life is left to them? Some years a go a journalist wrote that anyone who had ever visited a hospice comes away marvelling at the atmosphere of peace and dignity found there.

The downside of autonomy

The freedom to make our own decisions about our lives is a key principle in a free society. However, that freedom can only be exercised so long as it does not restrict the freedom of others to make their own decisions about their lives. (This is the central point in the debate about banning smoking in enclosed public places. I have a right to smoke and damage my own health. But do I have a right to smoke in places where tobacco smoke damages the health of other people?) Is there any evidence that lawful voluntary euthanasia restricts the freedom of others?

Voluntary euthanasia has been practised in the Netherlands for 20 years. Two studies, one in 1991 and later 2001, showed that there were between 2000 and 4000 cases of voluntary euthanasia each year in Holland; there were also about 1000 people each year whose lives were ended by their doctors not because they had requested it but because other people thought that their lives were intolerable and it would be better to relieve their suffering by ending their lives. What was voluntary for some people has become involuntary for others. When voluntary euthanasia was considered in Britain in 1994, the House of Lords Select Committee rejected it for this reason. The committee accepted the right of every person to refuse medical treatment but concluded that if voluntary euthanasia became lawful it would threaten the weak, the vulnerable and those without capacity.

12.6 When should medical treatment be withheld or withdrawn?

Introduction

The arguments for and against voluntary euthanasia are fairly straightforward. The central questions are how effective modern terminal care is in controlling pain and other symptoms and how much weight should be given to human autonomy. However, a much more difficult area is that where medical treatment is either withheld or withdrawn. When this happens in people for whom such treatment is life sustaining, they die. Is this right? What is the difference between killing somebody when they ask you to do so, because their illness or disability is unbearable to them, and letting them die by withholding treatment? There are two main areas to consider here:

- the right to refuse treatment;

- making decisions for people who cannot make them for themselves.

The right to refuse treatment

A mentally competent adult has the right to refuse medical treatment. However much a doctor may insist that it is a person's best interests to have a particular treatment, that the benefits will outweigh any side-effects, that person still has the right to say 'thank you, but I do not want to have it'. A not uncommon example concerns Jehovah's Witnesses. It is against their religious principles to have a blood transfusion. They cannot be compelled to have one. One of the authors has had to let a Jehovah's Witness die when a blood transfusion would have been life saving.

A recent high profile case in the UK was that of 'Miss B'. She suffered from a condition whereby she was paralysed from the neck down and could only breathe with the aid of a ventilator. In her view her life was intolerable and she had no prospect of recovery. She asked the doctors looking after her to remove the ventilator. They declined. Miss B went to court. Dame Elizabeth Butler-Sloss, England's senior family judge, ruled that Miss B was being treated unlawfully, because she had a legal right to refuse the treatment that the doctors continued to impose on her. She was moved to another hospital, where she was disconnected from the ventilator and she died peacefully in her sleep.

It is lawfully possible to exercise this right in advance by using an 'advance decision' (sometimes colloquially known as a living will). Such a decision has to be made while the patient is mentally competent and refers to a later time, under circumstances the person defines specifically, when a specified treatment is proposed to be carried out or continued by a person providing health care for him. The person states in the advance decision that he or she does not want to have the treatment. However, it only applies if at the time the treatment is being proposed, the person lacks capacity to agree to it or refuse it. Under a valid advance decision a doctor does not incur liability by observing it.

Making decisions for people who cannot make them for themselves

What is to be done for people who cannot agree to or refuse treatment, because they are mentally incompetent or in legal terms 'lack capacity'? People lack capacity if they are unable to make a decision for themselves. People are unable to make a decision if they cannot

- understand the information relevant to the decision;

- retain that information;

- use or weigh that information as part of the process of making the decision or

- communicate the decision (whether by talking, using sign language or any other means).

Examples of people who lack capacity are babies, children or people with dementia or who are unconscious. Under such circumstances somebody else has to make the decision whether or not to withhold or withdraw medical treatment. The test that has to be used is what is known as 'best interests'; is it in this person's best interests to start treatment or withdraw the treatment already being used? The only person who may lawfully make that decision is the doctor looking after the person. Of course the doctor must consult other people in making that decision, such as the person's spouse, other relatives, those engaged in caring for them or someone with power of attorney. Usually, with time and care, these decisions are not difficult to make, but sometimes the courts are asked to intervene because somebody else with a proper interest in the person does not agree with the doctors' advice (as in the case of baby Charlotte Wyatt, discussed earlier in the chapter) or because the case is particularly difficult and the doctors want to be sure that they may lawfully withhold or withdraw treatment. Over the last 10 years several such cases have come before the courts and the judges have very helpfully clarified what 'best interests' means.

The case of Tony Bland was the first case in which this was done. It is worth looking at it in some detail as subsequent cases have followed the principles that the English Law Lords laid down. Tony Bland had been a victim of the Hillsborough football stadium disaster in Sheffield in 1989. He had been in a permanent vegetative state (PVS) for three years as a result of the injuries he had received. In PVS the higher centres of the brain are destroyed. People with this condition show sleep–wake patterns. They respond reflexly to stimulation so that, if you were to pinch a toe, for example, the foot is withdrawn. (This response does not involve the brain but operates reflexly through the spinal cord.) There is no evidence of cognitive function. They cannot swallow and have to be fed through a tube passed into the stomach through the nose. With expert nursing care people with PVS can go on like this for years. There are two key questions about them.

- Are they alive in any normal sense of the word? Certainly there is a body that breathes but the *widely accepted, contemporary* biological basis of being a person has been permanently destroyed.

- Is feeding them through a tube an *artificial* means of support?

The doctors and the hospital in which he was being cared for brought Tony Bland's case to the courts. These two key questions were addressed by the courts and a final judgement was given by the House of Lords in February 1993. The Law Lords did take the view that Tony Bland was not alive in any normal meaning of that word. They also accepted that feeding him via a tube was a form of medical treatment and that it was a futile treatment because it conferred no benefit on him. Indeed, there was no benefit that could be conferred on him. Following this ruling the feeding tube was removed and he died some days later.

Some people have been profoundly disturbed by this case for three reasons.

(a)   Hitherto death had been defined as either cessation of the activity of the heart and lungs or the brain stem. Bland was not dead in this sense.

(b)   Food and hydration are basic human needs and while someone is alive they ought not to be denied them.

(c)   Since the intention of stopping the feeding was that Bland would die, stopping the feeding amounted to euthanasia.

The counter-argument to these points is that in PVS there is no prospect of being in anything other than the PVS until the person actually dies. A function necessary to life has been taken over artificially: Bland could not receive and swallow food or drink. This condition was the result of earlier unsuccessful treatment for catastrophic brain damage and it is therefore reasonable to 'let nature take its course'. The Law Lords, in the Tony Bland case, really boiled it down to two questions. These two key questions must always be asked before withdrawing fluid and food. First, was there any possibility that the condition was recoverable? Second, could he take and swallow food and fluid when offered in a cup or a spoon? If the answer to either question had been 'yes' then food and fluid must have continued to be given, but if the answer to both questions was 'no' then his vital body system of swallowing had failed and he had no hope of recovery. It should therefore be stopped.

Questions

Who do you think should decide whether or not to withdraw treatment from an adult who lacks capacity?
Do you consider food and fluid to be medical treatment in cases such as that of Tony Bland?

These principles together with a third one are now applied to other cases in trying to determine their 'best interests'. The third principle is the balance of benefit and harm. In other words, will the treatment confer any benefit or will it simply prolong suffering? Where the treatment is judged not to confer any lasting benefit and would also cause more suffering, the decision now (as in the case of baby Charlotte Wyatt) is either to withhold or withdraw it.

What is the difference between agreeing to a request for voluntary euthanasia and letting a patient die by withdrawing treatment? Those in favour of voluntary euthanasia say there is no real distinction because the result is the same – the person dies. Those who are against voluntary euthanasia but in favour of withholding or withdrawing treatment under certain circumstances argue that while the outcome is the same the motives are different. On the one hand the motive is to relieve suffering by killing the person; on the other hand the motive simply is to relieve suffering, recognizing that any further treatment is futile and whatever is or is not done, the patient will die. An important principle for doctors and other health workers is that they have a primary duty to save life. Where that cannot be done they have an absolute duty to care and to relieve suffering. In the Netherlands and Belgium they have concluded that this duty, under circumstances prescribed by law, can extend to killing somebody. It remains to be seen whether or not more countries follow suit.

13 A code of ethics for biologists?

'Hope you have – er – a good holiday,' said Hermione . . . 'Oh, I will,' said Harry . . . 'They don't know we're not allowed to use magic at home.'

From *Harry Potter and the Philosopher's Stone*, J.K. Rowling (1997)

13.1 Introduction

In Chapter 1 we showed that in order for science to progress there must be trust that other scientists have not falsified their data. Individual scientists have a responsibility to the scientific community and to science itself. This is not to say that as people scientists are likely to have higher personal ethical standards than others. Indeed, our experience of the world shows us that scientists are neither more nor less likely to cheat on their partners, break the speed limit or to fail to declare all their income to the tax inspector than anyone else. Scientists are not claiming a status as paragons of ethical virtue. However, integrity in the presentation of results is built into the ethos of science and those who fail to observe it pay the penalty.

We also showed in Chapter 1 that science is woven into the fabric of modern society; it is an activity to which society, *via* its decision-makers, allocates considerable resources against many other claims on those resources. At the time of writing (in January 2005) this is beautifully illustrated by the arrival, at considerable expense, of a 'moon-lander' on Titan, one of the moons of Saturn, after a seven-year journey in the parent space-craft. So, given that society supports the scientific enterprise, do scientists have wider responsibilities than just to science itself? Although in times past, the lone self-funded scientist may have been able to answer 'No', today, the answer is surely 'Yes'.

Introduction to Bioethics, by John Bryant, Linda Baggott la Velle and John Searle
Copyright © 2005 by John Wiley & Sons, Ltd.

13.2 The wider responsibilities of a scientist

One of us, in running training programmes for new post-graduate students, has suggested that in any research project a number of stakeholders may be identified. To a greater or lesser extent, the research scientist has responsibilities to all these stakeholders. The following list, which is not exhaustive, provides examples of the stakeholders and of the issues that arise in considering responsibilities:

Stakeholder	Issues
Scholarship/science itself	Falsification of data
Research community	Plagiarism
Immediate colleagues	Need to respect them and their work
Human subjects of research (e.g. in clinical trials)	Openness about aims and methods of research. No exposure to unacceptable risks
Funding agencies, such as research councils, charities and commercial organizations	Openness in report writing. Ownership and use of data
Employers	Direction of research. Ownership and use of data. Possible conflicts of interest
Wider society	Knowledge about how the results will be used; where appropriate, involvement in decision-making about applications of the work

In addition to these human stakeholders we can identify other areas of responsibility. These include

- use of animals in research: the need for a rigorous analysis of likely suffering versus scientific and medical benefits and the aims of refinement, reduction and replacement (see Chapter 4)

- possible environmental damage: if there is a possibility of this, then again there must a rigorous cost–benefit analysis.

In looking at this list of responsibilities, it will be obvious to our readers that sometimes responsibilities to different stakeholders may conflict with each other. Consider the following.

Dilemmas

1   *You work for a company and discover that it wishes to patent the sequence of a gene that has been your main subject of research. You believe that gene sequences should not be patented and further, that if this one is patented it will increase the costs of genetic tests. How should you proceed?*

2   *The pharmaceutical company for which you work is carrying out clinical trials with a new drug that has great promise for treatment of hypertension (high blood pressure). In parallel ongoing work in the laboratory, you and your research team discover that in a particular strain of laboratory rat there are rare but potentially serious side-effects. These had not occurred in the strain of rat used previously. The company does not wish these new results to enter the public domain and continues with the clinical trials. What is your course of action in these circumstances?*

In considering both cases, add in the following factors:
    You are the only breadwinner in your family, supporting a partner and three children.
    There are no other organizations in your area that employ biomedical research scientists.

These are both hard, but not unrealistic, cases. Other clashes of commitment may not be as extreme as these, but the point is that ethical decision-making, whatever ethical system is being employed (see Chapter 2), may not be straightforward.

Particularly interesting then are suggestions that biological and biomedical scientists need a formalized code of conduct that would first provide a framework for professional behaviour, and second assist in resolving tensions of the type highlighted above.

13.3 Should there be an ethical code of practice for bioscientists?

In 2001, John Sulston, the Director of the Sanger Centre and of the UK's part of the Human Genome Project, suggested that scientists should be required to sign up to a clear professional code of conduct, a 'Hippocratic oath for scientists' as he called it. He believes that this will not only help scientists in the performance of their work but will also

- ease public distrust of scientists

- prevent conflicts of interest arising where research is exploited for profit (as in the cases presented above)

- protect scientists from discrimination by employers who might prefer the scientists under some circumstances to be 'economical with the truth'

Specific points in the proposed code included promises to 'cause no harm and to be wholly truthful in public pronouncements'.

Although most bioscientists saw no need for an *oath* as such, there has been a growing interest in setting up codes of practice for scientists; such a code would be the framework for scientific and professional conduct. Points that often arise in discussion of codes of practice include several that were highlighted above in our discussion of stakeholders. For example, Ray Spier of the University of Surrey has been quoted as saying that a code of conduct should include

- being honest and open with colleagues and the public

- reducing experimentation on animals

- safeguarding the environment

- not plagiarizing or deliberately sabotaging other scientists' work

A more formal code has been published by the Institute of Biology (IOB),[1] the organization that, amongst other things, has general oversight of career development and professional standards of biological scientists in the UK. It is required, under the terms of its charter, to uphold and maintain the standards of biological practice. Much of the code of practice relates to appropriate professional behaviour in general but there are also many clauses that

[1] *Code of Conduct and Guide on Ethical Practice* (2004) Institute of Biology, London (see www.iob.org).

deal with the practice of science and especially of the biosciences. These include the following.

Members (i.e. members of the IOB) should

- exercise the highest standards of personal integrity and attention to accuracy in all scientific investigations and in the recording of results

- be aware of the ethical and societal implications of their work

- use their skills to advance and safeguard the welfare of humanity

- refrain from one-sided arguments

- discount and counter exaggerated, ill formed or prejudiced statements

- advocate suitable precautions against the possible harmful side-effects of science and technology and bring such recommendations to the public's attention

- identify the hazards and assess the risks of scientific and technological activities and processes

- use their knowledge and experience to protect and improve the environment

- minimize adverse effects on the environment and use best environmental practice.

This code leaves no room for the 'backroom boy' (or girl) working in their laboratory in isolation from the rest of society. It clearly attaches societal values to science, just as we did in Chapter 1, but it actually goes further than this. The code suggests that an awareness of the ethical and societal issues arising from their science should lead scientists to have an active involvement in preventing harmful uses of science. Of course, there is plenty of room here for differences of opinion about what is harmful or unethical, as is obvious from the material presented in this book. Nevertheless, there is a clear indication that bioscientists cannot sit on their hands and claim that they only find the results while 'society' is responsible for the way they are used.

On the other side of the Atlantic too there has been an interest in codes of ethics for scientists. One interesting example is that developed by Nancy Jones, who is an experimental pathologist at Wake Forest University, North Carolina. Her code, which is reproduced in the Appendix at the end of this chapter, also places the practice of science firmly in wider society and outlines the responsibilities of scientists to the various stakeholders that we identified in Section 13.2. This code, which is one of the most detailed that has been produced, is specifically aimed at biomedical sciences and particularly biomedical research that is very goal oriented. It would need modification to accommodate other areas of biology and other types of research. Nevertheless, like the more general code produced by the Institute of Biology, it does provide a framework for the pursuit of ethical standards in the practice of science. So, we end the book as we started it, with a question.

Question

Should professional bioscientists be required to sign up to a professional code of ethical practice?

Appendix
A Code of Ethics for Bioscience[1]

Nancy L. Jones, PhD
Associate Professor of Pathology Wake Forest University School of Medicine

Abstract

Society has demanded growing ethical awareness as bioscience increasingly controls our future. While bioscience has responded by initiating training in research integrity, this is often minimalist. The scientific communities of bioscience should reinvigorate their concept of professionalism. 'A code of ethics for bioscience' is an attempt to codify the rich professional heritage of bioscience. The code begins with articulating the goal for biomedical research and the responsibilities associated with the freedom of exploration. The rest of the code is divided between the principles for the practice of science and the virtues of the scientists themselves. Bioscientists have implicit role obligations. The practice of science is governed by several principles: objectivity, questioning certitude, research freedom, research reproducibility, respect for subjects and normalization through the scientific community. Scientists aspire towards several virtues: duty, integrity, accountability, altruism, excellence and respect for colleagues. This code is submitted as a prototype to begin to frame the ethics discussions among individual scientists and communities of bioscience as well as for use in the training of future scientists. Codification of principles and virtues as a teaching tool formalizes the expectation of what an *ideal* professional scientist aims for and emphasizes assimilation and identification as a member of the scientific profession.

[1] Reproduced here by permission of Dr Nancy Jones and of the Center for Bioethics and Human Dignity (http://www.cbhd.org/), which will publish the code in a forthcoming book on human enhancement biotechnologies.

Why do we need a code of ethics for bioscience?

Medical practice and human subject research are influenced by the Hippocratic tradition. While much of bioscience has an embedded ethos, it has not been formalized as such. Bioscience is defined as 'bench' or laboratory research in the fields of biology and biomedical sciences, rather than clinical research on human subjects. The activities of bioscience and technology are essential to provide solutions for our future, for both individuals and society. However, the public is often ambivalent in its attitude towards bioscience, drawn between aspiration and trepidation about where the activity of bioscience is leading. The influence is not unidirectional; the culture of science is equally influenced by our culture. Among the powerful influences moulding the culture of science are commercialization, globalization, and ownership of intellectual property. Formalization of a code of ethics, akin to a professional oath, would provide both internal and external ideals to shape the progress for bioscience.

Bioscience ethics code

The goal of biomedical research is the pursuit of knowledge in the biological and biomedical sciences with the ultimate goal of advancing human health and welfare of all human beings regardless of age or state of development. This pursuit should respect human life and dignity, remembering that science is a tool, a means to an end and never an end in itself. Underlying the freedom granted by society to pursue knowledge in the biological and biomedical sciences is the fundamental principle of trust in the integrity of scientists and the practice of science. In granting the privilege of freedom of inquiry, society assumes that scientists will accept the responsibility to act on behalf of the interests of all people. Scientists and the scientific community should accept the responsibility for the consequences of their explorations by guiding society in the development of safeguards necessary to judiciously anticipate and minimize harms.

Principles of the practice of science

Objectivity

The *prima facie* principle for the practice of science is objectivity. Objectivity is dealing with facts without distortion by personal feelings, prejudices or interpretations. The principle of objectivity is embedded in the ability to accept data that disproves a theory or hypothesis as readily as data that supports a theory or hypothesis. Scientists should strive to be objective in the experimental design, analysis and conclusions of their work. However, the very process of observation and interpretation of facts is a human and social venture and true objectivity is impossible. Scientists should endeavour to recognize the limitations their methods have for finding knowledge and be sensitive to the bias inherent in scientific activity. Scientists should be aware of biases introduced by external social and philosophical influences on the activity of science. Scientists should be open and honest about their relationships (e.g. involving their employer and/or funding mechanisms for the research), their commercial interests and/or the philosophical/political implications of the research, as these all can potentially influence the objectivity of scientists.

Questioning certitude

Questioning certitude is the readiness to question the current authoritative view or dogma in order to continue the process of advancing new knowledge. This principle asserts that no theory or fact is sacred; rather, experiments and conclusions should be continually re-evaluated in light of further discovery. The principle of questioning certitude helps to minimize bias in knowledge uncovered by any one study or field of study and acknowledges that there are inherent limitations of knowledge for scientific inquiry and interpretation.

Research freedom

Research freedom allows ideas to flourish within the scientific community with the understanding that eventually wrong concepts will be proven as such. Placing too much restriction on new ideas may prevent advances in knowledge. This freedom is granted within the community of science. However, forwarding these untested hypotheses or ideas as fact or as conventional scientific wisdom within the public domain is prohibited. Research freedom is not limitless; the practice of science does not condone unethical means of moving knowledge forward.

Research reproducibility

Quality scientific research should be able to be re-proven and to provide the groundwork for further exploration by any qualified scientist. Scientists should value the principle of open research to maximize the advancement of knowledge. Scientists should conduct their research in a way that allows open and thorough evaluation – as well as enabling repetition – of their research. When scientists are given privileged communication of research findings prior to public distribution of this knowledge (e.g. for purposes of evaluation for publication or funding) this knowledge must be kept as a sacred trust and not used until public distribution.

Respect for subjects

Scientists should uphold the highest ethical standards that respect all living things, with profound respect granted to human life and dignity. Respect entails designing experiments with the least invasive and destructive methods possible and avoiding unnecessary duplication of experiments. Respect necessitates designing experiments to answer the most pressing problems of humanity with stewardship towards limited resources.

The highest ethical standards for human subject research are codified in the Nuremberg Code (1946–49), the Belmont Report (1979) and the Declaration of Helsinki (1964, amended in 2000).[2] The duty of scientists includes protection of life, health, privacy and dignity of the research subjects. The scientific question must be of significant importance for human welfare and health and the well being of the subject must take precedence over the interests of science and society. Human subject research should when possible have prior animal experimentation showing a promising result with minimized risk and no other methods available for the same end. The risk should not be greater than the humanitarian importance of the problem to be solved, and no experiments are allowed with an *a priori* reason to expect death. The benefits gained from the research should be available to all populations in which the research is performed and no segment of the population should be excluded or bear the brunt of the experimentation.

Animal experimentation should have a peer-reviewed scientific rationale for the purpose and proposed use, justification of the species and number needed, and assurances that there are no other less-invasive or non-animal alternatives to answer the experimental question. Scientists should minimize suffering and harm to animals. Scientists should also be responsible for the welfare of animals and organisms using appropriate sedation, analgesia and anaesthesia and timely intervention to euthanatize suffering animals.

---

[2] These reports can be found on the web. Nuremberg Code, http://researchethics.mc.duke.edu/clinethics2.nsf/webpages/the+nuremburg+code; Belmont Report, http://ohrp.osophs.dhhs.gov/humansubjects/guidance/belmont.htm; Declaration of Helsinki, http://www.wma.net/e/policy/17-c_e.html

Community of science

The scientific community has been given an authoritative voice by society for esoteric knowledge in the domain of biological and biomedical sciences, and the skills thereof. With this authority, they bear the responsibility as the guardian for the integrity of science. While each individual member has been given freedom to pursue knowledge, the scientific community has the obligation to provide the normative processes for research activity through peer evaluation. The scientific community should provide proof of the veracity of individual findings through peer review and reproducing experimental results. The scientific community should afford a stamp of reliability only when other members of the scientific community can reconfirm the research. The scientific community should contextualize individual studies and provide assurances of the accuracy, the scope of the finding, and candid assessment of potential biases, conflicts of interests, and uncertainty of the knowledge. The scientific community has an obligation to correct inflation of an individual study's conclusion, misrepresentation of conventional scientific wisdom, or misuse of knowledge beyond the sphere of biological or biomedical science's ability to answer. The scientific community has an obligation to rebuke fraud.

The scientific community has the responsibility for training and accrediting future scientists in the practice of science. Students of science should be trained in both the knowledge and the philosophy of scientific practice.

Virtues of the scientist

Duty

Scientists should recognize the special status afforded to them by society as authorities on esoteric knowledge in the domain of biological and biomedical sciences. Scientists are considered agents seeking to uncover empirical objective knowledge or 'truth' in this sphere. As such, scientists should commit themselves to serve and guard humanity, including its individual members in both present and future generations. Biomedical scientists should accurately communicate and educate on the current understanding and uncertainty of their sphere of knowledge, seek to advance scientific understanding and respect the truth.

Integrity

Scientists should strive to be objective, fair, truthful and accurate. Scientists should speak publicly as authorities only about areas in which they have expertise. Integrity demands that research results are reported with as much objectivity as possible with no deliberate bias. Scientists should strive to present research in such a way as to avoid its possible misuse and misapplication.

Accountability

Scientists are accountable to their profession and society. Scientists have a duty to participate in the community of science to ensure that their scientific contributions, and those of their collaborators, are thorough, accurate and unbiased in design, implementation and presentation. Scientists are accountable for their public comments on scientific matters, which should be made with care and precision and devoid of unsubstantiated, exaggerated or premature statements. Scientists should seek to understand, anticipate and be forthcoming about the potential consequences (both benefits and harms) of their work.

Altruism

The scientist's primary focus should be on the best interests of humanity and not self-interest, commercial interests or the promotion of the industry of science. Care must be taken to assure that personal ambition and aspirations, or the desire to acquire profit or notoriety, does not influence professional scientific judgment. Scientists are obligated to be forthcoming with potential relationships that may pose a conflict of interest or influence their objectivity. The community of science provides the normative processes to ensure that personal ambition or potential conflicts of interest do not influence the objectivity of reviewed research.

Excellence

Scientists are committed to a lifestyle of learning and teaching. Scientists should remain current with developments in their field and transmit their knowledge to future generations of scientists. Scientists should share ideas and information and give due credit to the contributions of others. Scientists should keep accurate and complete laboratory records.

Respect

Scientists should treat associates and trainees with respect, regardless of the level of their formal education, encourage them, learn with them, share ideas honestly and give credit for their contributions. Scientists should credit colleagues, sources and published work for inspiration of their ideas. Scientists have responsibility for the health and welfare of their employees and trainees. Scientists should seek to minimize any potential risks in their laboratory work, informing their employees and trainees of these risks.

Glossary

Abortion termination of an established pregnancy; deliberately induced miscarriage.

Acrosome sac of proteolytic enzymes covering the nucleus of the sperm head. The enzymes are released at the point of fertilization and play a part in assisting the sperm to pass through the outer vestments of the egg.

Allergen a substance that induces allergy.

Anthropocentrism an attitude or approach to the environment centred on humankind; the rest of nature is regarded as being there for the good of humans.

Biocentrism an attitude to the environment centred on the biosphere; in respect of human activity in the world it means recognizing that we are part of the natural order and that we are one species out of many.

Biodiversity the range of living organisms in a particular habitat or community or in the whole biosphere.

Bioethics the ethics related to biology and medicine and to medical and biological research. After natively, the informing of ethics by biological knowledge.

Bioinformatics study of biological systems using the tools of information technology.

Blastocyst an early embryo at the point of implantation in the uterus. It consists of hundreds of cells of two main types: the inner cell mass, which goes on to become the embryo itself, and a surrounding layer of cells, the trophectoderm, which goes on to become the extra-embryonic membranes, including the placenta.

Blastomere undifferentiated cell of an embryo during the early cleavage stage.

Bovine spongiform encephalopathy (BSE, 'mad cow disease') a degenerative disease of the central nervous system of cattle, which is thought to be transmissible to humans; related to Creutzfeld-Jacob disease.

Carbon dioxide sink an organism, community or ecosystem that takes up carbon dioxide from the atmosphere.

Carcinogen substance that induces the formation of malignant tumours.

Chromosome one of a number of structures in the nucleus of a cell containing DNA (one DNA molecule per chromosome), the number of which is characteristic for the species. Each chromosome is, in effect, a subset of the total number of genes possessed by the organism.

Introduction to Bioethics, by John Bryant, Linda Baggott la Velle and John Searle
Copyright © 2005 by John Wiley & Sons, Ltd.

Clone organisms identical in genetic make-up, produced asexually from one stock or ancestor. May also be used as the verb, to clone. Both noun and verb are also used in the context of DNA, as in 'molecular cloning', the multiplication of particular of particular DNA sequences by 'growing' them in genetically modified bacterial cells.

Consciousness the state of being awake and aware of one's surroundings and identity; the totality of a person's thoughts and feelings.

Consequentialist line of argument that follows directly as a result of forgoing events or factors. Ethical system based on evaluation of the consequences of an action.

Creutzfeld-Jacob disease a degenerative, incurable disease of the human central nervous system, related to BSE (see above).

Deontology the study of duty and/or obligation as an ethical concept.

DNA vectors carriers of DNA sequences into the genomes of other organisms, e.g. bacterial plasmids.

Ecocentrism an approach to the environment that is centred on ecosystems. It values both the living and non-living components of ecosystems, seeing the place of humans within the total environment.

Ecosystem a biological community of interacting organisms and their physical environment.

Embryo the stage of life between the first cell division after fertilization until the completion of organogenesis. In the human this is during the first eight weeks of pregnancy.

Embryologist scientist who studies early development; scientist who manipulates human gametes and embryos under license in fertility treatment.

Ethics study of sets of moral principles; systematization of the principles involved in moral decision making.

Eugenics the process of improving the human population by genetic selection for desirable inherited characteristics.

Exogenous genes literally, genes from outside; in general this refers to genes that are transferred from one species to another.

F1 the first filial generation; hybrids arising from a first cross. Successive generations are denoted F2, F3 etc. P1 denotes the parents of the F1 generation, P2 the grandparents etc.

Fallopian tubes the tubes along which eggs (whether fertilized or not) pass into the uterus (womb).

Fetus/foetus the stage of development between organogenesis and birth.

Gamete a reproductive cell (spermatozoon or ovum) containing half the number of chromosomes of a somatic cell and able to unite with one from the opposite sex to form a new individual.

Gene an individual hereditary unit in a chromosome consisting of a characteristic sequence of DNA.

Gene flow the spreading of genes resulting from out-crossing and from subsequent crossing within a population.

Genome the total genetic content of an organism.

Genomics the study of the genomes of organisms, particularly in relation to information content and sequence organization.

Germ cell a cell belonging to the specialized cell lineage that gives rise to gametes (sperm and eggs) in a multicellular animal or plant.

Heterozygous having dissimilar, alternative forms of a gene for a given characteristic.

Homozygous having identical genes for a given characteristic.

Huntington's disease hereditary, incurable and degenerative disease of the human central nervous system.

Implantation process by which a embryo becomes embedded in the lining of the womb.

In vitro Latin for 'in glass'. When used as *in vitro* fertilization, it refers to laboratory techniques in which sperm, eggs and embryos are cultured outside the body.

Inbreeding breeding through a succession of parents belonging to the same stock/lineage, or within very closely related lineages.

Infertile permanently and completely unable to have children. Affects both sexes approximately equally with a range of causes.

Intra-cytoplasmic sperm injection (ICSI) micromanipulation technique in which a single immobilized sperm is injected directly into the interior of the egg.

Introgression the introduction, by plant breeding, into elite crop lines/varieties of genes/traits from related varieties (or even from closely related wild species). Has also been used recently to describe (the possibility of) transfer of transgenes (see below) from GM crops into wild populations.

Modernism the reliance on science and technology as the basis for reason. It has its origins in the 18th century.

Morals the study of goodness or badness of human character or behaviour, or of the distinction between right and wrong; the outworkings in action of systems of ethics.

Mutant an organism that has undergone a permanent, heritable change in structure and/or function based on a change – mutation – in DNA.

Natural law an ethical system based on (what are perceived to be) the laws of nature or on consideration of the natural function of an object or being.

Nematodes invertebrate phylum of round worms.

Oncogene gene that makes a cell cancerous; typically a mutant form of a normal gene (proto-oncogene) involved in the control of cell growth or division.

Oocyte female gamete; egg, ovum.

Out-crossing breeding with another population or species.

Penetrance the frequency, measured as a percentage, with which a gene shows any effect.

Personhood the quality or condition of being an individual person (in the *Concise Oxford Dictionary*, a person is defined as a human being).

Plasmid small circular DNA molecule that replicates independently of the genome. Used extensively as a vector for transfer and cloning of DNA.

Pluripotent capable of forming many different types of cell (see also *totipotent*).

Polymorphism occurrence within a population of an organism of different varieties based on different alleles of a given gene.

Post-modernism a philosophy that rejects modernism and any ideas of universal principles, truths and meta-narrative. Embraces a pluralistic and often relativistic approach to ideas and ethics.

Promoter the tract of DNA adjacent to a gene that contains the gene's 'on–off' switch.

Rational egoism an ethical system in which actions are evaluated only according to the consequences for oneself.

Rationality possessing the faculty that transcends understanding and provides *a priori* principles: sensible conduct – to do what is right or practicable.

Recombinant DNA novel combination of DNA sequences made by recombining different DNA sequences in the laboratory, e.g. insertion of a mammalian gene into a bacterial plasmid.

Semen seminal fluid containing spermatozoa.

Sentience the power of perception by the senses.

Somatic concerned with the body cells (as opposed to the germinal cells).

Speciesism a form of prejudice against members of other species because of their species.

Subfertile having reduced fertility, either temporarily or permanently. Can often be helped by artificial reproductive technologies and other forms of treatment. Affects both sexes approximately equally with a range of causes.

Superovulation artificial stimulation of the ovarian cycle with hormones to induce formation of multiple mature eggs.

Syngamy the coming together of the genetic complement of two gametes after fertilization.

Theocentrism literally 'God-centredness'; in environmental ethics an approach to nature based on the idea that it belongs to God.

Totipotent genetically capable of forming a whole organism (including the placenta).

Transgenes genes that have been transferred via a vector from one organism to another.

Utilitarianism the doctrine that actions are right if they are useful or for the benefit of the majority; guiding principle of the greatest happiness for the greatest number.

Virtue ethics an ethical system based on deciding what is the most virtuous action. It is centred on the concept of the virtuous character; it is held that the practice of virtue will lead to a person becoming more virtuous.

Zygote cell formed by the union of spermatozoon and oocyte; a fertilized egg.

Suggested Further Reading

Armstrong, S. J. and Botzler, R. G. (eds) (2003) *The Animal Ethics Reader*. Routledge, Abingdon.

Attfield, R. (2003) *Environmental Ethics*. Polity, Cambridge.

Baggott, L. M. (1997) *Human Reproduction*. Cambridge University Press, Cambridge.

Beauchamp, T. L. and Childress, J. F. (1989) *Principles of Biomedical Ethics*. Oxford University Press, Oxford.

Bruce, D. (2002) Finding a balance over precaution. *Journal of Agricultural and Environmental Ethics* 15, 7–6.

Bryant, J., Baggott la Velle, L. and Searle, J. (eds) (2002) *Bioethics for Scientists*. Wiley, Chichester, UK.

Bryant, J. and Searle, J. (2004) *Life in Our Hands*. IVP, Leicester.

Carson, R. (1962), *Silent Spring*, Houghton Mifflin, New York.

Centre for Bioscience (2004) *Bioethics Briefings*, The Higher Education Academy, Centre for Bioscience, Leeds (*Bioethics Briefings* are available online at the Centre's website – see below).

Cook, G. (2004) *Genetically Modified Language*, Routledge, Abingdon.

Deane-Drummond, C. E. (2004) *The Ethics of Nature*, Blackwell, Oxford.

Dolins, F. L. (ed) (1999) *Attitudes to Animals: Views in Animal Welfare*. Cambridge University Press, Cambridge.

Food Ethics Council (2003) *Engineering Nutrition: GM Crops for Global Justice?* Food Ethics Council, Brighton.

Fukuyama, F. (2002), *Our Posthuman Future*, Farrar, Strauss and Giroux, New York, and Profile, London.

Harris, J. (1998) *Clones, Genes and Immortality*, Oxford University Press, Oxford.

Harris, J. and Holm, S. (eds) (1998) *The Future of Human Reproduction: Ethics, Choice and Regulation*, Oxford University Press, Oxford.

Holland, S. (2003) *Bioethics*, Polity, Cambridge.

Hursthouse R. (2000) *Ethics, Humans and Other Animals*. Routledge, Abingdon.

Jochemsen, H. (ed) (2004) *Human Stem Cells: Source of Hope and of Controversy*. Prof. dr. G.A. Lindeboom Institute and Business Ethics Center, Jerusalem.

Introduction to Bioethics, by John Bryant, Linda Baggott la Velle and John Searle
Copyright © 2005 by John Wiley & Sons, Ltd.

Jonsen, A. R. (1998) *The Birth of Bioethics*. Oxford University Press, New York.

Kuhse, H. and Singer, P. (eds) (1999) *Bioethics: an Anthology*, Blackwell, Oxford.

Light, A. and Rolston, H. (eds) (2003) *Environmental Ethics: an Anthology*. Blackwell, Oxford.

Marteau, T. and Richards, M. (eds) (1996) *The Troubled Helix*, Cambridge University Press, Cambridge.

McConnell, R. L. and Abel, D. C. (2002) *Environmental Issues*, Prentice-Hall, Upper Saddle River, NJ.

Mepham, B. (1995) *Food Ethics*, Taylor and Francis, Abingdon.

Moore, P. (2001) *Babel's Shadow: Genetic Technologies in a Fracturing Society*, Lion, Oxford.

Nuffield Council on Bioethics (1999) *Genetically Modified Crops; the Ethical and Social Issues*, Nuffield Council on Bioethics, London.

Nuffield Council on Bioethics (2002a) *The Ethics of Patenting DNA*, Nuffield Council on Bioethics, London.

Nuffield Council on Bioethics (2002b) *Genetics and Human Behaviour: the Ethical Context*. Nuffield Council on Bioethics, London.

Regan, T. (2004) *The Case for Animal Rights*, University of California Press, Berkeley, CA.

Robinson, D. and Garrett, C. (1996) *Ethics for Beginners*, Icon, Cambridge.

Shakespeare, T. (1998) Choices and rights: eugenics, genetics and disability equality, *Disability and Society* 13, 665–682.

Shamoo, A. E. and Resnik, D. B. (2003) *Responsible Conduct of Research*, Oxford University Press, New York.

Shiva, V. (1998) *Biopiracy*, Green, Dartington, UK.

Singer, P. (1986), *Animal Liberation: a New Ethic for our Treatment of Animals*, Cape, London.

Singer, P. (1994) *Rethinking Life and Death*, Oxford University Press, Oxford.

Singer, P. (1996) *Animal Liberation*, Pimlico, London.

Snowden, R. and Snowden, E. (1993) *The Gift of a Child: a Guide to Donor Insemination*. University of Exeter Press, Exeter.

Stock, G. (2002), *Redesigning Humans: Choosing our Children's Genes*. Profile, London.

Teichman, J. (1996), *Social Ethics*, Blackwell, Oxford.

Warnock, M. (1998) *An Intelligent Person's Guide to Ethics*, Duckworth, London.

Warnock, M. (2003) *Making Babies: Is There a Right to Have Children?* Oxford University Press, Oxford.

Wilmut, I., Schnieke, A. E., McWhir, J., Kind, A. J. and Campbell, K. H. S. (1997), Viable offspring derived from fetal and adult mammalian cells. *Nature* 385, 810–813.

A Small Selection of Useful Websites[1]

Abortion: Roc versus Wade: members.aol.com/abtrbng/410us113.htm

Altweb: alternatives to animal testing on the Web: www.altweb.jhsph.edu/education/history,htm

American Society of Plant Biologists: www.aspb.org

British Broadcasting Corporation: www.bbc.co.uk

Center for Bioethics and Human Dignity: www.cbhd.org

Centre for Bioethics and Public Policy: www.bioethics.ac.uk

Christian Medical Fellowship: www.cmf.org.uk

Crop Biotech Net/The International Service for the Acquisition of Agri-biotech Applications: www.isaaa.org

Department for Environment, Food and Rural Affairs, UK: www.defra.gov.uk

Euthanasia: www.euthanasia.cc and www.euthanasia.org

Food Ethics Council: www.foodethicscouncil.org

Google search engine: www.google.co.uk and www.google.com

Higher Education Academy Centre for Bioscience: www.bioscience.heacademy.ac.uk

Human Fertilisation and Embryology Authority: www.hfea.gov.uk

Human Genome and other Genome Projects: www.doegenomes.org/

Ethical, Social and Legal Issues relating to the Human Genome Project: www.ornl.gov/sci/techresources/Human_Genome/elsi/elsi.shtml

Institute of Biology: www.iob.org

Irish Council for Bioethics: www.bioethics.ie

Network for European Woman's Rights: reproductive rights: www.newr.bham.ac.uk/pdfs/Reproductive/biblio.rep.art.pdf

Newspapers: the websites of the 'serious' newspapers in all relevant countries carry archives of all news items including those with a bioethical content

Nuffield Council on Bioethics: www.nuffieldbioethics.org

Research Defence Society: www.rds-online.org.uk

Roslin Institute: www.roslin.ac.uk

Sanger Centre: www.sanger.ac.uk

Stanford Encyclopaedia of Philosophy: the moral status of animals: www.plato.stanford.edu/entries/moral-animal/

Wellcome Trust: www.wellcome.ac.uk

---

[1] There are many others!

Index

Introduction to Bioethics, by John Bryant, Linda Baggott la Velle and John Searle
Copyright © 2005 by John Wiley & Sons, Ltd.